Usability Evaluation in Industry

Usability Evaluation
in Industry

EDITED BY

PATRICK W. JORDAN
Philips Corporate Design

BRUCE THOMAS
Philips Corporate Design

BERNARD A. WEERDMEESTER
Nederlandse Vereniging voor Ergonomie

AND

IAN L. McCLELLAND
Philips Corporate Design

Taylor & Francis
Publishers since 1798

UK Taylor & Francis Ltd, 1 Gunpowder Square, London, EC4A 3DE
USA Taylor & Francis Inc., 1900 Frost Road, Suite 101, Bristol, PA 19007

British Library Cataloguing in Publication Data

A catalogue record for this book is available from the British Library.

ISBN 0 7484 0314 0 (cased)
ISBN 0 7484 0460 0 (paperback)

Library of Congress Cataloguing in Publication data are available

Cover design by Hans Jacobs and Robin Uleman (Philips Corporate Design).

Typeset in Times 10/12pt by MHL Typesetting Ltd, Coventry.

Printed in Great Britain by T.J. Press (Padstow) Ltd.

Contents

About the Editors

Pat Jordan is a human factors specialist with Philips Corporate Design, where he is the leader of two large research projects. These investigate pleasure in product use and usability evaluation methods. Previously he was employed by the University of Glasgow in the Department of Psychology – first as a research assistant and then as a lecturer. He has been published 30 times in academic and technical journals, and books of conference proceedings. He presented and co-wrote the video *User Interface Performance Measurement* and co-authored an instructional CD-ROM on usability to be released in 1996. Dr Jordan is a regular speaker at international conferences and was General Chair of the seminar 'Usability Evaluation in Industry' on which this book is based.

Bruce Thomas studied psychology and worked briefly for a consultancy in the UK before moving to TüV Rheinland in Germany. There he worked on research, development and consultancy projects, including workplaces in power stations, hotel and restaurant kitchens and electronic equipment for cars. He spent a year on secondment to the University of Bremen investigating workplaces on ships. He has worked at Philips Corporate Design since 1990, where he is primarily responsible for coordinating human factors support for the development of communications products.

Bernard A. Weerdmeester graduated from the Agricultural University in Wageningen in 1982. After graduation he joined the Dutch PTT, where he carried out research on human factors in telecommunications. In 1986 he went to the University of Twente where he worked as a researcher and teacher of ergonomics. He returned to the PTT in 1989 and was head of the user interface group. With two former PTT colleagues he started up 'Raakvlak', a consultancy bureau on user interfaces. After a year, however, he left Raakvlak and started 'Usable', a bureau on the same market. Since 1987 Bernard Weerdmeester has been president of the

Dutch Ergonomics Society (Nederlandse Vereniging voor Ergonomie). He is co-author of some books on ergonomics, including *Vademecum Ergonomie* (in Dutch) and its English version, *Ergonomics for Beginners*.

Ian L. McClelland first trained as a mechanical engineer and then moved to Loughborough University of Technology in 1970 where he completed a post-graduate course in ergonomics. In 1972 he received his master's degree and joined the Institute for Consumer Ergonomics where he undertook a wide variety of research and consultancy work for government, business and industry. In 1986 he joined Philips Corporate Design as manager of the Applied Ergonomics Group. His interests include the integration of usability engineering principles into the process of user interface design, the involvement of users as 'partners' in interaction design, and evaluation methodology.

Contributors

Jans Aasman
KPN Research, PO Box 421, 2260 AK Leidschendam, The Netherlands

Chris Baber
School of Manufacturing and Mechanical Engineering, University of Birmingham, Edgbaston, Birmingham B15 2TT, UK

Nick I. Beagley
Centre for Human Sciences, DRA, Farnborough GU14 6TD, UK

Hans Botman
Utrecht Institute for Higher Education, School for Communication Studies, Kaap Horndreef 10, 3563 AS Utrecht, The Netherlands

John Brooke
Redhatch Consulting Ltd, 12 Beaconsfield Way, Earley, Reading RG6 2UX, UK

Gary Davis
Gary Davis Associates Ltd, Ergonomics Design Consultants, Baldock, Herts, UK

Edward A. Edgerton
University of Paisley, Department of Applied Social Studies, Paisley, Scotland

Ian A.R. Galer
ICE Ergonomics, 75 Swingbridge Road, Loughborough, Leics, UK

Tedde van Gelderen
Philips Corporate Design, Building OAN, PO Box 218, 5600 MD Eindhoven, The Netherlands

Mark Hartevelt
Philips Corporate Design, Building OAN, PO Box 218, 5600 MD Eindhoven, The Netherlands

Reinoud Hulzebosch
Informaat, PO Box 5, 1200 AA Hilversum, The Netherlands

Anthony Jameson
Universität des Saarlands, PO Box 151150, D-66041 Saarbrücken, Germany

Graham I. Johnson
Cognitive Engineering, Technology Development Group, AT&T Global Information Solutions (Scotland) Ltd, FP&S, Kingsway West, Dundee DD2 3XX, Scotland, UK

Patrick W. Jordan
Philips Corporate Design, Building OAN, PO Box 218, 5600 MD Eindhoven, The Netherlands

J.A.M. (Hans) Kemp
Philips Corporate Design, Building OAN, PO Box 218, 5600 MD Eindhoven, The Netherlands

Jurek Kirakowski
Human Factors Research Group, University College, Cork, Ireland

Jan Legters
Department of Human Factors and Psychology, Hoogovens IJmuiden, PO Box 10 000, 1970 CA IJmuiden, The Netherlands

Ian L. McClelland
Philips Corporate Design, Building OAN, PO Box 218, 5600 MD Eindhoven, The Netherlands

Miles Macleod
Andersen Consulting, 2 Arundel Street, London, WC2R 3LT

Enid M. Mante
Social Science Research Department (ITB), PTT Telecom, The Netherlands

Bert Mulder
Centre for Interaction Design, Utrecht School of the Arts, Faculty of Art, Media and Technology, PO Box 2471, 1200 CL Hilversum, The Netherlands

Marjolein van Nieuwkasteele
Philips Sound and Vision, 5600 MD Eindhoven, The Netherlands

Ron Oosterholt
Philips Corporate Design, Building OAN, PO Box 218, 5600 MD Eindhoven, The Netherlands

Magdalen Page
ICE Ergonomics, 75 Swingbridge Road, Loughborough, Leics, UK

Dick Rijken
Centre for Interaction Design, Utrecht School of the Arts, Media and Technology, PO Box 2471, 1200 CL, Hilversum, The Netherlands

Karma M.A. Sierts
Social Science Research Department (ITB), PTT Telecom, The Netherlands

Elselien Smit
Brace BV, Calandstraat 18-G, 3016 CB Rotterdam, The Netherlands

Robert B. Stammers
Psychology Group, Aston University, Birmingham B4 7ET, UK

Neville A. Stanton
Department of Psychology, University of Southampton, Highfield, Southampton SO17 1BJ, UK

Bruce Thomas
Philips Corporate Design, Building OAN, PO Box 218, 5600 MD Eindhoven, The Netherlands

Arnold P.O.S. Vermeeren
Delft University of Technology, Faculty of Industrial Design Engineering, Product and Systems Ergonomics Group, Jaffalaan 9, 2628 BX Delft, The Netherlands

Edwin van Vianen
Philips Corporate Design, Building OAN, PO Box 218, 5600 MD Eindhoven, The Netherlands

Govert de Vries
Philips Corporate Design, Building OAN, PO Box 218, 5600 MD Eindhoven, The Netherlands

Bernard A. Weerdmeester
Nederlandse Vereniging voor Ergonomie, Plein 564, 1054 SJ Amsterdam, The Netherlands

Preface

As consumers become more and more sophisticated in terms of what they demand of products, so the role of the human factors specialist is becoming ever more challenging and demanding. Whilst users may once have seen ease of use as a bonus, in the case of many products it is now often taken for granted. In order to offer more, those involved with designing for usability now have to look beyond ease of use to consider issues such as the emotional and hedonic benefits associated with product use.

Philips Corporate Design is committed to excellence in design for usability, and human factors input is incorporated throughout the design process. We see activities such as the seminar 'Usability Evaluation in Industry', which brought together 25 of Europe's leading human factors practitioners and researchers, as an important forum for the exchange of ideas and views amongst those at the top of the profession. These professionals had the opportunity to learn from each other and to take the benefits of these shared ideas and approaches back to their own companies or research establishments. Philips Corporate Design was therefore happy to support the two-day event, both financially and through the commitment of the staff involved in the organization and running of the event.

The chapters in this book give an overview of the material presented at the seminar as well as the discussion that ensued. It should prove essential reading to those involved in the practice of human factors and all those involved in research to support usability issues.

Dr Stefano Marzano
Senior Director, Philips Corporate Design
Eindhoven

Acknowledgements

This book is based on the international seminar 'Usability Evaluation in Industry' that was held in Eindhoven, The Netherlands, on 14 and 15 September 1994. Financial support for the seminar came largely from Philips Corporate Design. We are grateful to them for supporting the idea of the seminar and giving us the necessary funding to put the idea into practice.

We would also like to thank the Dutch Ergonomics Society (Nederlandse Vereniging voor Ergonomie) for giving their support through endorsement of the seminar.

Finally, of course, we must thank the delegates for the chapters that make up this book and for two days of useful and stimulating discussion. In addition to those who contributed to the book, Kate Bentley of PTT Research and Gerard van der Heiden of Rabobank attended the seminar and made valuable contributions through their presentations and discussion.

Pat Jordan
Eindhoven

CHAPTER ONE

Introduction

PATRICK W. JORDAN, IAN L. McCLELLAND and BRUCE THOMAS

Philips Corporate Design, 5600 MD Eindhoven, The Netherlands

Issues relating to usability are increasingly coming to prominence. This is reflected in, for example, the ever-growing literature relating to usability issues. This includes academic journals such as *Behaviour and Information Technology* and *Human–Computer Interaction*, books such as Nielson's *Usability Engineering* and Wiklund's *Usability in Practice*, and a host of newspaper and magazine articles. Similarly, there are now many national and international conferences and other gatherings which deal with usability issues. These include Computer–Human Interaction (CHI), the Human Factors and Ergonomics Society Conference and, of course, the seminar 'Usability Evaluation in Industry', which formed the basis for this book. Perhaps, though, the most significant reflections of how seriously usability issues are taken are the increasing number of professional human factors specialists employed in industry and the ongoing research efforts of academia in this area.

As the products that we use at home and in our workplaces become ever more complex in terms of features and functionality, it becomes vital that those involved in the design of these products consider the needs and limitations of those who will be using them. If these are not taken into account, products that are created with the intention of delivering some benefit can end up being more trouble than they are worth. Users are becoming more sophisticated with respect to their expectations about product performance. These include their expectations about a product's usability. Users seem no longer willing to accept having to struggle with products as the price to pay in return for 'technical wizardry'. Whilst once usability may have been seen as a bonus, it is rapidly becoming an expectation, with users becoming disenchanted with products which do not support an adequate quality of use. In the end these products will also disenchant those who are manufacturing them as they will find that their customers start looking elsewhere. Usability is becoming established as an important issue with respect to the marketing and sales of products, thus it can have significant commercial implications.

Manufacturers are continually seeking new ways of trying to gain a competitive edge. In the current business environment, the opportunity to make competitive gains in 'traditional' ways – such as by high reliability or reduction in manufacturing costs – is being continually eroded. A major reason for this is the sophistication of manufacturing processes, the high quality and consistency of which make it difficult for any one manufacturer to gain a significant advantage over another in these respects. Exploring the advantages of user-centred design is seen by some manufacturers as a way of enhancing possibilities to gain competitive advantage.

Most often products which are difficult to use may result in annoyance and frustration – for example, with video cassette recorders (VCRs) that are difficult to program, computer applications that we can't understand, or central heating systems that switch themselves on or off when we least expect. However, usability can also be associated with more serious issues, in particular the level of risk to personal safety in using a product.

Consider, for example, the case of a driver using in-car navigation equipment. It may take him or her, say, ten seconds to interpret navigation advice presented by a system. During this time the attention of the driver will be diverted from what's happening on the road. This is likely to make it more difficult to deal with anything unexpected that may happen – for example, a car stopping ahead or a child running out into the road. Similarly, with emergency products, such as fire extinguishers, it is vital that any member of the public can pick up the product and use it at the first time of trying. Again, this means that such products have to be carefully matched to users' expectations.

The importance of designing usable products is now receiving increasing attention from industry. Usability is coming to be seen as part of industry's responsibility to its customers, and, of course, it is also becoming important in terms of selling products. More and more products are advertised as being 'user-friendly' or 'ergonomically designed' and where once potential customers saw usability as a bonus it is increasingly becoming an expectation.

So how do these industry-based human factors specialists go about ensuring that the products that their organizations produce are usable? This book is a collection of chapters from many specialists working in industry as well as academics who work closely with industry. The chapters describe a number of techniques used in the evaluation of products – techniques designed to give an overview of the suitability of the product for the user and to feed back information that can be used to improve the quality of the design. The authors also relate their experiences of applying their techniques in an industrial setting, where they are working under pressures and constraints which may be considerably sharper than and certainly different from those experienced in academia where the majority of the techniques were originally developed. The contributors to this volume met together for a two-day seminar in Eindhoven, The Netherlands, in September 1994, from which the chapters included in this book emerged. A number of issues also emerged during the course of the seminar discussions. These issues are addressed in Chapter 26, in which overall conclusions are drawn on the basis of what emerged from the discussions during the seminar.

This book should be of benefit to all those involved in product design and the evaluation of products for usability. Through their experiences the authors have gained clear ideas of what does and doesn't work under industrial constraints, and their chapters should be a valuable source of advice and support to those working under the same pressures. This can, perhaps, be seen as a contrast with more academic writings – such as those contained in journals or conference proceedings – where the emphasis may be more on scientific validity and procedural rigour, rather than on the pragmatics of achieving a useful result as quickly and cheaply as possible.

This book should also prove valuable to academia. Research aimed at the support of usability evaluation is increasingly being carried out at universities and other research establishments throughout the industrialized world. Often, however, there may be a gap between what academia provides and what industry finds helpful. A contributory factor to this may well be a lack of understanding of industrial practices and pressures. Again, material relating to this may be difficult to get hold of in journals and conference proceedings. Certainly, many academics will work with individual companies and may gain an understanding of how some industry-based ergonomists work. However, the experiences related in this book should give a broader overview, demonstrating the practice of industry-based usability evaluation as a whole.

PART ONE

Elements of Usability Evaluation

CHAPTER TWO

A combined effort in the standardization of user interface testing

E. VAN VIANEN

Philips Corporate Design, Eindhoven, The Netherlands

B. THOMAS

Philips Corporate Design, Eindhoven, The Netherlands

M. VAN NIEUWKASTEELE

Philips Sound and Vision, Eindhoven, The Netherlands

INTRODUCTION

Philips, being a multinational consumer electronics company, needs to conduct studies of the usability of its products with users in a variety of countries, settings, etc. This means that investigations, or parts of them, must be conducted by local agencies. One of the problems at the moment is that for individual projects (depending on, for example, the market research agency used), the approaches, analysis of results, reporting, etc. are often different. This makes it impossible to compare the results of tests done on interfaces for different TV sets, for example. Since there is a drive for continuous improvement within Philips, we need to be able to use previously developed interfaces as benchmarks.

This chapter presents some of the results of a joint project between the market research department of one of Philips' Business Groups (BG), Television (TV), and the Applied Ergonomics group of Philips Corporate Design (PCD). Since this has been a joint project, some of the techniques described are traditional human factors techniques, whereas others are derived from market testing.

The goal of this project has been to develop a standardized approach to the evaluation of user interfaces for BG, taking a very pragmatic approach. The end product was to be a document describing which techniques to use when, what kind of questionnaire to use, etc. This document will serve not only as an internal checklist and selection instrument, but also as a set of 'rules' to be followed when an outside contractor is employed. The issues to be addressed were defined at the start of the project:

- purpose of test,
- specification of subjects,
- design of user/usability tests,
- measures to be taken,
- questionnaires used,
- interview questions,
- tasks to be executed,
- presentation of results.

In the following, excerpts from the original document resulting from this project (which is still in progress) are presented and briefly discussed. To help structure the approach, the methods and techniques are linked to the product creation process (PCP) of this specific BG. The PCP, which is predetermined by BG TV, consists of the following main phases:

- Know-how phase: Aims at gathering, generating and applying know-how in several fields.
- Concept phase: Encompasses carrying out work for future product generations.
- After product range start (PRS): The phase leading up to the specification of a rough product range.
- After commercial release (CR): The product is released for delivery and distribution on the market.

The term 'test' is used in this chapter to mean any involvement of users in getting feedback on the user interface of a television. The terminology used here may not always be in line with 'standards' used within the human factors community, but corresponds to that used throughout Philips.

PURPOSES OF USER INTERFACE TESTS

Generic purposes

The purpose of a test will influence the methods and techniques to be used, and what is to be measured. Hence it is of great importance to define the purpose of the test before starting. Some fundamental and generic questions that can/must be addressed regarding the purpose are for example:

Are we trying to find out what people want with regard to user interfaces?
A starting point can be to determine the customer's needs and wishes, which can be translated into a user requirements specification.

Are we trying to find out what people accept with regard to user interfaces?
Taking a given system, the goal of a test might be to determine the acceptance of such a system by our customers.

Are we trying to find out what people are able to use?
Another goal might be to test to what extent consumers are able to use, independent of the aesthetic qualities, a system they are confronted with, e.g. whether there are problems with particular interactions or interaction mechanisms.

Specific purposes related to the PCP

Although the purpose of user interface (UI) tests has to be fine tuned with each new test for each new project, some more specific purposes can be distinguished, based on the phase in the PCP in which the test will be conducted. Some examples for each phase are:

Know-how phase
Testing to generate new ideas for concepts
Testing to select the first basic ideas and concepts
Inventory of existing problems
Testing to answer specific questions

Concept phase
Testing to answer specific questions
Testing to determine whether acceptance criteria have been met
Testing to select ideas and concepts
Testing to identify usability problems

After PRS
Testing the final concept in detail, and verification of implementation

After CR
Inventory of problems and remarks
Confirmation of usability criteria
Detailing of problems

SPECIFICATION OF THE PARTICIPANTS

Before each test, a detailed specification of the participants has to be made based on the product brief (with reference to the target group(s), etc.). The idea is to make use of a checklist in which the main characteristics of participants are listed. Before starting a test those responsible have to go through this list to specify the participants needed. In principle, much of this information can be found in a User Requirement Specification. Examples of the checklist items for participants are:

- *Number of participants.* This is dependent on: goal of the test, the technique used, and the statistical reliability (depending on the confidence limits set).
- *Gender.* This is dependent on the target group(s), but in principle a 50/50 split.
- *Age.* This is dependent on the target group, and the upper/lower limits.
- *Ownership* (according to desired profile) of TVs, audio equipment, VCRs and/or satellite tuners.

TOOLS AND TECHNIQUES FOR UI TESTS

As in the purpose of the test we can also link the design of the test to the PCP. In general, in the know-how phase the emphasis will be on qualitative techniques, and in other parts of the PCP on the more quantitative. Table 2.1 shows for each phase of the PCP, examples of the purposes of the tests (as described above) linked with the appropriate techniques. The techniques are selected on the basis of suitability, and prior experience. In this chapter the term 'technique' is used to mean the framework within which the user interface test is done. Of course a combination of techniques can also be used during one test. Examples of techniques are user workshops, focus groups (see Jordan, 1994), informal usability tests (see Thomas, Chapter 12; Vermeeren, Chapter 14), and expert appraisal (based on Ravden and Johnson, 1989; see also Johnson, Chapter 20).

Related to selection according to purpose is the phase in the PCP at which a test is carried out. Table 2.2 shows a selection of techniques that can be used in specific

Table 2.1   Overview of testing techniques linked to their purpose and the phase in the PCP

PCP phase	Purpose	Techniques
Know-how	Generation of new concepts	■ User workshops ■ Focus groups
	Inventory of problems	■ All techniques (see Table 2.2)
Concept	Answer specific question	■ All techniques (related to a specific question)
	Compliance with acceptance criteria	■ Usability tests
PRS	Testing final concept	■ Usability tests
CR	Inventory of problems	■ Inventory
	Confirmation of usability criteria	■ Usability tests

Table 2.2 Overview of testing techniques and the tools and the equipment that can be used (depending on the phase in the PCP)

PCP phase	Technique	Description	Tools	Equipment
Know-how	User workshop	A group of users is put together to discuss more generic issues in relation to product(s), is asked to prioritize functions, re-design specific parts, etc.	■ Discussions ■ Interviews ■ Questionnaires ■ Co-design ■ Task execution ■ Brainstorming	■ Product(s) ■ Simulation (paper or SW)
	Focus group	A group of users is put together to discuss particular issues related to a product (simulation)	■ Discussions ■ Questionnaires	■ Product(s) or simulation(s)
	Usability test	Formal test in which respondents perform tasks and in which effectiveness, efficiency and satisfaction are measured	■ Task execution ■ Questionnaires ■ Interviews ■ Thinking aloud protocol ■ Observation	■ Simulation(s) or product(s)
Concept	Informal test	Short test in which internal people perform tasks, perhaps together with face-value test, etc.	■ Task execution ■ Questionnaires ■ Interviews ■ Thinking aloud protocol ■ Observation	■ Product(s) or simulation(s)
	Usability test	See above	See above	■ Elaborate simulation
PRS	Usability test	See above	See above	■ Physical prototype
CR	Inventory of usage	Finding out what functions are used, etc.	■ Data logging	■ End product(s)
	Usability test	See above	See above	
	Inventory comments/remarks	Finding out what problems consumers have with the final product	■ Interviews ■ Questionnaires ■ Incident diaries	■ Product(s)

Table 2.3 Overview of the strengths and weaknesses of some testing techniques

Technique	Strengths	Weaknesses
User workshop	User involvement in finding solutions Intense and effective initial concept generation Assesses appropriateness of target group and defines it in more detail Combine evaluation/solution generation	Much effort required to set up More expensive initially, but provides more in-depth information at an earlier phase
Focus group	Can discover the unexpected Revealing and prioritizing issues that are important to users Rapid information gain from large number of users	Poor basis for decision making Information mainly from opinion leaders Little measurable data
Usability test	Can measure task performance User involvement	More expensive, but provides more in-depth information at an earlier phase More time consuming, but can save discussion time
Informal test	Quick	Measurements can be unreliable Users often not from target group
Inventory of usage	Objective information Easy to process	Need for special equipment
Inventory of comments/ remarks	Input from users based on end product	Self-selective Dependent on motivation

phases of the PCP, together with short descriptions, an overview of the tools that can be used and the equipment needed. The latter can include the 'product' to be tested, as well as the internal tools for data logging that are part of the tool box.

As with all testing techniques, they all have their own strengths and weaknesses (see Table 2.3). Understanding the strengths and weaknesses of particular techniques is important in tailoring the choice of methods to the particular questions raised. So, for example, the discussion in a focus group might be influenced by an opinion leader in the group.

MEASURES AND OUTPUTS

When looking at the techniques used, some statements can be made with regard to their output, again in relation to the phase of the PCP. Table 2.4 gives an overview

Table 2.4 Overview of testing techniques, what they are used to measure, and the output they generate

Tools	Measures	Output
Discussion	Qualitative information	■ User needs and wishes ■ Acceptability of concepts ■ First selection of concepts ■ Not detailed design directions
Questionnaires	Rating scales ranking	■ Measure of satisfaction ■ First selection of concepts ■ Weighting of problems ■ Weighting of functions
Interviews	Qualitative information	■ Identification of problems ■ Selection of concepts ■ Acceptability of concepts ■ Identification of needs and wishes ■ Patterns of use
Co-design	Qualitative information	■ Design ideas
Checklist	Qualitative and quantitative information	■ Identification of strengths and weaknesses ■ Recommendations ■ Scores on human factors variables
Task execution	Time on task Percentage success	■ Usability criteria ■ Identification of problems
Thinking aloud protocol	Qualitative information	■ Identification of problems ■ Identification of strategies ■ Identification of user expectations
Observation	Qualitative information	■ Description of tasks ■ Descriptions of environment/context of use
Data logging	Number of key presses Frequency of use	■ Description of strategies used ■ Inventory of errors/problems

of the tools, the measures used, and the output that those measures can generate. The table lists only those measures and outputs that are relevant for us, so it is by no means exhaustive.

SCENARIOS FOR TESTING

Part of the standardization of procedures also concerns the way the tests have to be set up with respect to the order of events, and the content of each event. Again the goal is to ensure that in principle the same set-up is used for the same techniques, resulting in comparable data. Short examples of what we call scenarios for a user workshop and a usability test are as follows:

The basic scenario for a *user workshop* includes the following events/activities:

- welcome
- introduction of all participants
- explanation of the goal, and presentation of agenda
- introduction of problem(s)/product(s)/issue(s)/concept(s)
- exploration (playing with product(s)/concept(s), discussion on problem(s)/issue(s))
- formalizing/visualizing results
- creative session (drawing, concept generation, etc.)
- evaluation of results of the creative session via discussions/exploration
- wrap-up.

In principle, the scenario for a *usability test* includes the following:

- exploration
- task execution
- interview
- questionnaire.

Here, 'exploration' means a period of time in which a participant is free to try out the menu system of the TV set, look at the manual, try out the remote control, etc. During this exploration phase the participants are asked to think aloud (i.e. give comments, remarks, etc.). The investigator for example has to:

- note down all remarks
- note down what functions the participant uses, looks up in the manual, etc.
- note down what strategy the participant uses when going through the menu system and the manual
- make sure that the participant looks at all the parts of the menu system
- ask questions at specific points
- note down the kinds of mistake the participant makes
- note down spontaneous behaviour.

The administration of the questionnaires and interviews, and what questions should be asked, are described in the next section.

QUESTIONNAIRES AND INTERVIEWS

This section describes the content and the administration of interview questions and questionnaires, using user workshops and usability tests as examples.

User workshops

Principle   The user workshop makes use of the cross-fertilization of ideas in a group of participants. Such a group is more active and more self-creating than, for example, a focus group. Less attention is paid to proposed concepts and more to the creation of new concepts.

Set-up   The funnel method (starting with exploring an issue in its entirety, and ending up focusing on specific details) is applied here, in order to get group members to think seriously about the subject and to base their reactions on reality. Various means can be used to help to inspire creativity, including:

- making drawings,
- making collages,
- 'pro and con' discussions,
- imagining 'storybook' scenarios , such as living on another planet or in another age.

When participants are asked to be creative in finding new solutions, etc., it is important that what is being asked is clearly defined (in chunks). Thus respondents should not be asked to come up with a new UI concept for overall control, but to give their ideas for representing information concerning sound settings on the screen, for example.

Usability tests

Principle   A formal test in which users are required to perform specific tasks and in which efficiency, effectiveness and satisfaction (according to ISO 9241) are measured according to, for example, percentage success, the time taken to complete a task, questionnaires, etc.

Set-up   Depending on the purpose of the test, the questions relate to:

- on-screen menus and their operation
- use of the remote control
- use of the manual.

We now present some examples of the questions that are used in the interviews and questionnaires. The contents of these questions are partially based on the Software Usability Measurement Inventory (SUMI) list (see Kirakowski, Chapter 19) and the System Usability Scale (SUS) from DEC (Brooke, Chapter 21).
  Examples of generic questions for interviews include:

What is your general opinion about the operation of this product?
What are the worst aspects? (for operation, remote control and the manual)
Did you have any problems executing the tasks? If so, what problems?

Did you find any parts confusing or difficult to understand?

Were any parts particularly irritating although they didn't cause major problems?

What were the most common mistakes you made when using the set?

What kinds of improvement would you like to make? (for operation, remote control and the manual).

Examples of generic questions for questionnaires include:

Perceived effectiveness:

It was a problem to ... (activate/operate a specific function, find something in the manual).

Perceived efficiency:

It took a long time to ... (activate/operate a specific function, find something in the manual).

I made a lot of errors when ... (activate/operate a specific function, find something in the manual).

Satisfaction:

I found carrying out ... (activate/operate a specific function, find something in the manual) satisfying.

I found using the 'remote control X' satisfying.

Examples of questions to determine the overall usability of the product include:

I would like to use 'product X' frequently

I think that 'product X' is more complex than necessary

I found 'product X' fun to use

I would imagine that most people could learn to use 'product X' quickly

I found 'product X' cumbersome to use

I felt confident using 'remote control X'

I needed to learn a lot of things before I could get going with 'product X'

I found that the information in the manual was not always clear

CONCLUDING REMARKS

This project is a good example of cooperation between different disciplines which in a lot of companies might be in competition with each other since they are both responsible for 'user testing'. By setting aside our own interests, we have achieved a mutual respect and willingness to cooperate, and thereby provided an example of synergy.

Another important outcome has been the agreement within Philips BG itself. Although not formalized, it is the intention of the authors to make the final document part of the official PCP. The result is that everyone who wants to perform user tests of any kind will be obliged to comply with the guidelines.

We are now working on a project in which we are making use of the first results of the standardization project, especially with regard to which techniques to use, questions to administer, etc. We are also finding that although these guidelines were developed specifically for BG TV, they are beginning to help in other areas. For example, the principles outlined here are being used to guide the set-up of a series of tests for communication products in the Netherlands and France. Needless to say, the principle of standardization of user interface tests is an important issue for the entire Human Factors group at PCD and the Marketing Research and Support Department at BG TV.

REFERENCES

JORDAN, P.W. (1994) Focus groups in usability evaluation and requirements capture: A case study, in: S.A. Robson (Ed.), *Contemporary Ergonomics 1994*, London: Taylor & Francis, pp. 449–453.

RAVDEN, S.J. and JOHNSON, G.I. (1989) *Evaluating the Usability of Human–Computer Interfaces: A Practical Method*, Chichester: Ellis Horwood.

CHAPTER THREE

Usable usability evaluation: If the mountain won't come to Mohammed, Mohammed must go to the mountain

ELSELIEN SMIT

BRACE B.V., 3016 CB Rotterdam, The Netherlands

INTRODUCTION: THE USABILITY REALITY

As an interaction and usability practitioner I have often been involved in usability evaluations that have given me the unpleasant feeling that the work has been a waste of time and money. This usually happens with evaluation requests for already developed (pre-)market versions of applications. The clients, being proud of their user-friendly product, see the usability evaluation as the finishing touch in the optimization of the application, and as a confirmation of their 'user-centred' approach. However, the evaluation often shows that large improvements are necessary, even in the information structure. In the (pre-)market stage it is usually too late to do this. The result is that all people involved in the project are disappointed. The application designers are disappointed (and/or angry) because the evaluator may tell them that many things should have been designed differently. The team may have worked for a long time to produce the application, and a usability evaluator, looking around for just a few days, is able to run down the application completely. The willingness to accept such evaluation results is low. In my role as usability evaluator, I have also been disappointed, because only a few recommendations for improvement are executed, usually those concerning the information presentation, to camouflage the real problems. The client is also frustrated; the product appears to have many deficiencies even though it was produced by a team with a high level of expertise. Yet there have been no real complaints from users. Were the shortcomings real or was the evaluator simply a perfectionist? The client is also frustrated in that he has had to pay for a complete set of recommendations, only a part of which was useful at this stage.

One reason for all these frustrations is the lack of insight into usability engineering, on the part of the client as well that of the development team. A consequence is that each person involved has different expectations about the evaluation goals and results. Therefore usability evaluators must concentrate not only on the usability of the application, but also on the usability of the evaluation. We always preach that the conceptual model of an application must be logical and clear for the users, to enable them to build a correct mental model. We must apply the same principles to clarify our usability approach to people with an information technology based mental model. Explaining why learnability and fault tolerance are important does not work – such concepts are too abstract. To construct a conceptual model for an application, we use the application domain and knowledge about users and their tasks. In the same way we must use the knowledge and goals of the client and developers, to enable them to develop a correct mental model of what usability is, and why it is important. This is a precondition for achieving a positive attitude towards usability in a development environment.

A USABLE USABILITY METHOD

On the basis of the experiences described above, a usability evaluation method has been developed that takes the mental model of the client and the development team as a starting point. The idea is simple. The evaluator begins to talk with these people about the application and its environment of use, about their aims with the application, and about critical success factors. Then you must find out which requirements must be met to reach these aims, and soon you will find yourself in the middle of a usability discussion. The challenge is to reach the point at which people begin to see that usability is not something that can be stuck on or taken away from an application, like a ribbon around a present. Usability does not work like that; it is an integral part of the application and its information environment. If you reach this point you are ready to execute the usability evaluation. There is a beginning of an awareness of what usability means for this particular application, and because you show your involvement in their problems, they are also more likely to come to trust the evaluator. Both are important for a successful evaluation.

AIMS AND STEPS

The aims of the evaluation method are to approach usability with the mental model of the client and the development team as a starting point, and to reach acceptance of the need for usability improvements by increasing the understanding of, and motivation for good usability. The method contains the following steps:

1   Specify the usability evaluation request of the client. Verify the aims of the client, developers, buyers and the users of the application.
2   Define the usability requirements. Define the requirements that must be met to reach the aims.

3   Prioritize requirements and confirm the selection of evaluation method. List
     the requirements and determine priorities and points of attention for the
     usability evaluation.
4   Prepare and execute the usability test. Make an expert assessment and/or
     execute test(s) with the (future) users of the application.
5   Analyze and report the results and present them to the client.

HOW DOES IT WORK? THE METHOD APPLIED

The application of the method is illustrated in this chapter by means of a case study.
CTB Automation and Pluriform Partners, two Dutch companies, have developed an
information system for the building industry, called the Building Management
System (BMS). BMS is intended for companies such as building entrepreneurs and
building project organizations. BMS supports tasks like financial administration,
project planning, administration and management, stock management and
invoicing. The system consists of a set of modules that can be used separately or
in combination, depending on the needs of the user organization. Early versions of
the system had been tested by users, and their comments collected during contacts
with the user organizations. A systematic inventory of the usability of the system
had not yet been made. A few months before BMS was to be launched on the
market, CTB Automation asked for a usability evaluation.

Step 1: Specify the usability evaluation request of the client

The first thing to do was to find out more about the mental models of the client, the
development team and people involved in the sales and promotion of BMS. How
did they see the role of BMS in the building industry; what were their aims with
BMS; and what were critical success factors for the system? Further information
was needed about the building industry environment in which BMS and users
cooperated and performed their tasks, and about the aims of buyers, users, and
BMS itself. More insight was obtained by interviewing the client and a developer,
reading the user manual, taking a first glance at BMS, and by talking with people
who were in contact with users and the environment of use.

Mental models and aims of client and developers

For the client, the aim of BMS was to take over and support many tasks in the
building industry, and to become the information heart of an organization. The
client considered the fact that the system not only supported daily work processes,
but also gave detailed insight into all aspects of work processes to be one of its
strengths. This makes BMS a strong tool for process control and cost management.
Suppose an organization wished to know how many hours per month were spent
transporting materials to, from, and between workplaces. BMS answered such

questions by combining different kinds of information. The client realized that getting the most out of BMS required from many people a more precise way of working than they may have been used to. If the input was not up to date, the output would not be useful.

For the developers, the aim of BMS was to make a sophisticated, very flexible, object-oriented information system for the building industry. It had to be flexible in its ability to generate, combine, select and present all kinds of data very quickly, and in its capability to be adapted, to be tailored to the users' needs, and to be extended with new types of data, new layouts, new menu options, or new modules. What made BMS user-friendly was the fact that input, once given, could be used everywhere and also to generate new data where possible. For example, a user could book the costs of a purchase invoice on a project, and at the same time the invoice would be processed automatically in the accounts department. BMS was also user-friendly in the consistency of presentation of menu layers and levels of task forms, for instance, which were all named to avoid getting lost. Real user errors were, according to the developers, almost impossible because the system warned them immediately if the input was not acceptable.

The aim of CTB and Pluriform with BMS was to become market leaders in information systems for the building industry. This meant that they wanted to bring to the market a product that was more complete, more flexible and more user-friendly than those of their competitors. To encourage the use of BMS on a large scale, CTB had developed a distribution formula with a low threshold for buyers. The consequence of this formula was that CTB could only profit if the system were used intensively by the user organizations.

In discussing these aims *together* we came to the conclusion that a critical success factor for BMS was that the system should fit so well into an organization that it could be easily and widely adopted. BMS had to be so 'extrovert and expressive' that users could easily perceive its possibilities. Further, the behaviour of BMS had to be predictable and correspond to the users' expectations, leading them to trust the system and their own way of working with it. This was important to make users curious to explore more of the system, and to use it for as many purposes as they could find. BMS had to be so attractive that users would prefer it to sticky notes, pocket calculators, secret hand-written product lists, or paper invoice files. With this conclusion there was suddenly a common ground for involvement in the usability of BMS. This was where we really began to talk about usability.

Aims of buyers and users

To achieve the critical success factor, the system had to meet the aims of buyers and users. The buyers were the building industry organizations, whose activities are often spread over different locations. The planning, management and control of people, work and materials took take place at the headquarters, but important parts of the executive work were done at other workplaces. The aim of buyers was to have a system that would increase the effectiveness and efficiency of work processes. By integrating a large number of work processes (and work locations) in one system, BMS also had to improve communications between different departments and processes. Further, the buyers wanted to have up-to-date and directly accessible information available in order to provide a good service to customers. The success of BMS would be influenced not only by the involvement of each person in his or her own role in a work process, but also by the extent to which BMS fits into these processes and is able to maintain, or maybe even to increase, the involvement of users. Finally, the work processes and the organization themselves gave BMS a purpose and made its use meaningful.

The users of BMS could be divided roughly into two groups: those with technical backgrounds, and those with administrative backgrounds. Each group consisted of people with different levels of knowledge and expertise in their own professional domains as well as in working with computers. Their aims with the system were derived from their roles within the organization – work planner and calculator, project manager, book-keeper, or logistics employee – as well as from their personal aims and motivations for their own work and for the organization. A general aim of BMS users was to execute the tasks within their roles correctly and efficiently, while also working with pleasure. Pleasure can come not only from personal involvement in the task, a way of working, or good results, but also from the satisfaction of doing a job within a team, or from social contacts with colleagues. BMS would change existing ways of working, and thus could influence existing sources of job satisfaction. BMS may also have introduced new sources of satisfaction by being 'a good pal' to work with.

The evaluation request

When the context of the usability request has been clarified, the initial request of the client can be formulated more precisely.

First, a usability evaluation was drawn up for BMS taking into account the following general aims and success factors:

- Aims of buyers: to increase the effectiveness and efficiency of work processes and communications by improving and integrating them.
- Aims of users: to do their work correctly and efficiently, but also with pleasure.
- Aims of BMS: to manage, take over and/or support a large number of administrative tasks in work processes in the building industry, and to create new, relevant information by processing user inputs and available data in various ways. Another aim of BMS was to be flexible enough to be tailor-made for organizations, and capable of being extended in the future.
- Aims of CTB and Pluriform: to become market leaders in information systems for the building industry, and to realize this by offering a system that takes over and supports work processes in various fields of the building industry in an adequate, sophisticated and user-friendly way.

- Critical success factor: to offer a system that is so attractive in its functionality and usability that it would be used intensively (many hours per day by a large number of users).

Second, the evaluation was implemented by means of an expert assessment. The results were reported and recommendations were made for the improvement of BMS. It had to be taken into account that the system would be launched on the market within two months.

Step 2: Define usability requirements

Knowing what we were working for, the requirements that had to be met to reach the aims identified during step 1, had to be defined. Some examples:

- BMS had to have the potential to become a market leader in information systems for the building industry (it had to be stronger than its competitors).
- BMS had to be 'a good pal' to work with: it had to involve users in the work (not only ask for input and give feedback, but be a partner in a dialogue).
- BMS had to be extrovert and expressive: the work that BMS executes or supports had to be organized and represented as recognizable user tasks. Users had to be able to perceive which tasks they could execute within their functional role, and should not be bothered with tasks or information that did not belong to their role and were not accessible to them. The names of tasks or task symbols had to be recognizable to the users. BMS had also to invite users to explore its possibilities.
- BMS required a precise way of working (accurate input). It had to encourage users to work precisely by making the work easily accessible and executable without the need for extra attention to the system itself.
- The flexibility of BMS had to be perceivable to the users (for instance, the fact that once the costs of a purchase invoice were booked, the invoice was also processed by the financial administration).

The approach followed here did not necessarily result in new types of requirement. Most of them resembled general usability rules. The difference is that they were derived from the aims of this particular application and formulated for this particular situation. This increased the involvement and understanding of the client and the development team. When the evaluation showed that a requirement was not being met, it was clear that one of the aims that the client (not the evaluator) defined as important, would not be reached. This motivated people to implement changes.

The list of requirements also contained elements that could not be categorized purely as usability aspects. For instance, a requirement that BMS must be strong enough to become the information heart of an organization, had many facets – functional, organizational, motivational, infrastructural, as well as usability – in these cases there is always the problem of whether a judgement could be given in the context of a usability evaluation.

Step 3: Prioritize requirements and confirm the selection of evaluation method

The requirements were presented to the client. They showed what would be the focus of the evaluation study and thus helped to avoid unrealistic expectations. The overview also helped in the selection of requirements that received priority. For BMS, a decision was taken to bring the application on the market two months after the first appointment about the evaluation, so that changes had to be executable within this period. Given the critical success factor that BMS should invite users to use and explore the system, it was decided to give priority to everything necessary to achieve this. It was then up to the usability evaluator to translate the client's commercial aim into user-centred points of attention for evaluation. The usability matrix presented by Rijken and Mulder (see Chapter 6) may help to do this in a systematic way. It may also help to verify whether the initial choice of evaluation method is the most suitable one. Usually a first choice of a method is made during the specification of the request of the client. Clients want to know at an early stage what they can expect from a study, and how much time and money will be involved. But specific requirements may give reasons to change the initial idea, or to point at the restrictions of the initial choice. For instance, to find out whether BMS invited users to explore, it was thought best to test it with real users, and not just to make an expert assessment. At this stage the evaluator has arguments to explain why, and the client has a better insight on which to judge them. For CTB, the evaluation of BMS was their first experience with a usability specialist. They wanted to start with a small project to gain insight into the added value, and in this context, an expert assessment was preferred to a larger-scale evaluation with users.

The requirements discussion is also meant to decide how to judge whether or not a requirement has been met. If qualitative information is desired it can be agreed with the client to give a general impression about the extent to which each requirement has been met, list shortcomings, and formulate recommendations for improvement. If a quantitative judgement is desired the usability requirements can be turned into usability statements, and then tested. For example:

- A usability requirement for BMS: the names of tasks or task symbols must be recognizable to users.
- The usability statement: BMS has good usability if the names of tasks or task symbols are recognizable to users.
- A method of testing the usability statement: Ask 20 future users of BMS to list their daily tasks, and then ask them to show you if and where they can find these tasks in BMS.
- The criterion for accepting the usability statement: 99% of users must be able to recognize and find tasks incorporated in BMS.

- Report the findings with remarks and/or problems, and with recommendations for improvement.

Step 4: Prepare and execute the usability test

Once an agreement has been reached about the requirements to be tested, the usability test can be prepared and executed. What needs to be done depends on the method(s) chosen. For BMS the evaluation was executed by means of an expert assessment. The preparation consisted in formulating points of attention for the evaluation session and organizing the session itself. Some examples of points of attention were:

- Does BMS 'show' its possibilities? Do these correspond with user tasks and information needs?
- Make a list of a number of functional roles and tasks. Verify whether they are recognizable and easily accessible. Check also whether related tasks can be handled as such.
- Does BMS involve users in a dialogue, or does BMS only 'talk to' users?

The organization of the session was simple for this (one-person) expert assessment. The system and the user documentation were ready to use. A CTB employee, who was familiar with the system and its use, was standing by in case of problems, or to provide explanations as and when needed. The client was also available to give additional domain information where necessary. It is important to have an informant in the background (not too close): as soon as you begin to use the application, new questions will almost always arise.

Step 5: Analyze and report results and present them to the client

The findings of the evaluation must be analyzed and judged in relation to the usability requirements. It is good to step back first and give an overall impression of the usability of the application, rather than to concentrate on specific subjects, or only on the weak points. If the strong points are known, they can be used to improve or camouflage weaker ones.

On the basis of the judgement, recommendations for improving the usability must be formulated and reported. Think beyond the problem you want to solve. For instance, for BMS the evaluation made clear that the large number of possibilities offered by BMS might have led users to feel lost, which would work against the critical success factor. A short-term solution within the constraints was to categorize the menu options according to functional roles of employees, e.g. the options relevant to a book-keeper, a project manager, or a sales person. This reduced the number of possibilities for each user. But thinking in the longer term and in a wider context, we preferred another solution. We left the menu options, and explored a concept based on the functional roles of employees, and their tasks and cooperation in the organization. BMS was intelligent enough to understand different definitions of user roles, and to present matching tasks that can be manipulated directly, instead of going through a menu hierarchy with subtasks. A guide (which can later be developed into an intelligent agent) could point the user to more advanced options, or could suggest a next step.

Apart from the fact that many user organizations are (technically) not yet ready for this approach, this could not be realized before BMS was launched on the market. But with this aim in front of us, we could realize the framework of the concept, and make it work within the current restrictions (e.g. with a very small guide function). With a new release, the next step could be made on the migration path that has now been marked off for BMS. Working like this gives the stakeholders (including the evaluator) the satisfaction that the usability evaluation adds something to the application, and does not just rectify shortcomings.

The evaluation report usually only outlines such a solution. A discussion with the client and the development team is then necessary to judge the feasibility of and to obtain support for the idea, before it is worked out in detail. In general, such a discussion of the recommendations is indispensable to get the most out of the evaluation. It is necessary to explain why certain recommendations have been formulated, to discuss alternatives in the case of an infeasible recommendation, and to talk about priorities in the implementation of the recommendations. Keeping in touch with the development team is equally important during the next stage, when the recommendations are implemented. It is important to be certain that the recommendations are correctly interpreted, to monitor the coherence in the application, and to avoid the need for people to fall back on information technology habits to solve a problem.

THE ADDED VALUE OF THE APPROACH

The difference between the approach described here and the 'traditional' usability evaluation is that the client's situation is taken as a starting point, rather than the usability problem itself. By doing this the problem is treated in a recognizable context. The evaluator is required to leave the ivory usability tower and to show interest and involvement in the client's aims. But it is worth it, because you will receive trust in return. Trust is the basis for the added value of this approach: it increases involvement, improves usability awareness, and contributes to the acceptance of the evaluation results.

From a usability point of view one wonders whether this approach compromises the quality of an evaluation. Experience has taught me that this need not be true. In an organization without a breeding ground for usability, the contribution of a usability specialist will remain small. Usability is a supporting domain, but it can not survive if it is not supported itself. Thus, also trying to understand the context of the application development is almost a precondition for good results. Maybe the quality of evaluation will be different, but it will certainly not be diminished. The main thing is that usability practitioners must be willing to go to the mountain, for as long as the mountain does not come to them.

ACKNOWLEDGEMENTS

I wish to thank Mr W. van Doesburg from CTB Automation in Ede, The Netherlands, for the pleasant cooperation in the BMS evaluation project, and for his permission to use this experience for publication. I also wish to thank the development team of Pluriform Partners in Veghel, The Netherlands, for their positive collaboration in realizing the good usability of BMS.

CHAPTER FOUR

Combining physical, physiological and subjective data in the evaluation of product usability

IAN A.R. GALER and MAGDALEN PAGE

ICE Ergonomics, Loughborough, Leicestershire, UK

INTRODUCTION

The word 'product' is widely used and now refers to a great variety of artefacts, including consumer goods such as TV sets and cookers, industrial equipment, vehicles and their components, but also software systems, savings and investment plans, and health services. This broad range involves a rich mixture of human requirements and capacities and, whilst it is generally meaningful to think of the usability of any 'product', the methods used to assess this, the types of measures involved, and the commercial environment in which the assessment is made, are so diverse as to render impractical any attempt at generalization, at least insofar as evaluative methods are concerned. This chapter therefore focuses primarily, though not exclusively, on the usability evaluation of three-dimensional devices intended for use in the home, at work, or in private or public vehicles. They are, generally, products which are grasped or manipulated and which require the application of some force or effort in order to use them. The idea of 'usability' in this context is dominated by ease of use, comfort, and acceptability rather than by issues of, say, logical transparency which might indicate the usability of a software product. We will, however, make reference to one example of the latter.

Ergonomists and product designers have attempted to measure the usability of products for many years. The methods employed have ranged from simple, informal appraisal (one designer to another: 'do you reckon that's about the right size?') to large-scale user trials with tight experimental protocols. They are applied to initial models through to marketed products, and to product components as well as the complete product. The measures used may yield qualitative or quantitative data: from the summarized results of a group discussion to a statistically interpretable measure of time to complete a task, or the number of errors made.

The evaluation method chosen is, of course, a reflection of the precision required and the resources available, as well as the *type* of information that is sought. As an illustration, we recently wanted to assess the effort or strain involved in the use of a computer input device. Our favoured method was to use electromyography (EMG) to measure the level of activity in the forearm muscles under different conditions of use of the device. This would have yielded numerical data that were a direct reflection of the corresponding human activity. However, it would have been very costly and time-consuming to use EMG techniques, and because this was only a pilot, exploratory exercise we concluded that our requirements for precision, and our resources, could not justify such an approach. We therefore resorted to subjective methods, using comfort/discomfort as the primary measure. Under the circumstances, this served our needs adequately.

One of the problems with evaluation methods is in deciding precisely what it is we are trying to measure, and satisfying ourselves that two different ways of measuring what appears to be the same thing yield broadly similar results. Psychometric testing experts have been tackling these issues for many years: in intelligence testing, for example, not only is a mutually agreed definition of intelligence required (no easy matter) but different measures of this concept should correlate. (Cynics have defined 'intelligence' as that which is measured by intelligence tests; perhaps 'usability' is what is measured by usability evaluation methods!) We have encountered analogous problems in product evaluation, even within a given class of measures: for example, we have used the subjective method of preferential choice in evaluating the acceptability of different designs of car instrument displays. Asking a panel of subjects 'which of these five designs do you find most attractive?' gives one answer; a superficially innocuous change of wording to 'which of these five designs would you prefer to have in your own car?' gives a quite different answer. This example illustrates not only the slippery nature of concepts such as acceptability, ease of use, and usability, but also the methodological pitfalls which can trip the unwary evaluator and misdirect the client.

It is an attractive, and reasonable, proposition that different measures of what appears to be the same human attribute or experience, should agree with each other; that the (subjective) perceived effort of using a product should correspond with the (objective) level of muscle activity, and with the (objective) amount of force applied. However, this seems difficult to achieve. Over the past three years, for example, we have undertaken detailed research on seating comfort, one aspect of which has been to see whether seat comfort (or, more accurately, discomfort) measured subjectively corresponds to a measure of the pressure over the area of contact between the seat surface and the body of the sitter. The ultimate purpose of such an exercise was to try to develop a way of predicting long-term seat comfort from a single, rapid and standardized measure of seat pressure distribution rather than a lengthy, expensive and imprecise series of subjective user trials. We have been able to get some way towards this goal, but it has been very difficult to achieve the degree of agreement and correlation we would wish; the correlation is sensitive to minor changes in the experimental protocol and makes us less than optimistic about the prospect of developing a robust and practical tool.

We may fail to find a strong correlation between two variables for a number of reasons. The most obvious is that there is in fact little or no relationship between them: the two variables are indeed measuring different things and there is no *a priori* reason why they should correlate. This clearly has serious implications for usability measures for as long as we maintain the view that usability is, broadly, a unitary concept which should be reflected in a range of potential indices. Discarding the concept means that a great deal of attention will be needed to identify the exact meaning of the term for a particular product in a particular context. Another reason for the absence of correlation, however, is that the variables may be related but in a nonlinear fashion. This has not been evident in our own studies although it remains a possibility. A nonlinear relationship should be readily discernible through exploratory data analysis and the development of a nonlinear statistical model. The scientific implications of such a model are interesting, although we have yet to explore this. A third reason is that either or both of the variables does not vary sufficiently, by accident or design, for a correlation to be observed. It is self-evident that such variation is necessary, and although analytical means do exist to overcome the problem to some extent, these do have their limitations. We have encountered this issue in some of our own studies, and this is described below. A final reason for the lack of correlation is that error masks the relationship. Again, we have encountered this particularly in the use of subjective measures and we discuss this in more detail below.

The grail of co-relating measures of the same characteristic or experience, especially if these measures are at different levels on the reductive scale, is nonetheless attractive. It generates a richer picture of the object of measurement, and mutual agreement between measures gives us a greater confidence in the validity and reliability of each – especially when, subsequently, it may for resource reasons be necessary to rely on only one measure in the evaluation of a specific product. And it must be acknowledged as attractive for a client to be able to advertise or justify the benefits of his product at more than one level ('not only will you find X more comfortable than its competitors, but tests have shown that it actually needs less effort to use!'). Over the course of the past six years we have undertaken a series of projects and commissions in product evaluation which have involved us in the simultaneous collection of data of a number of different types and levels which are all focused on the same human characteristic or experience. The next section describes some aspects of this work.

RESEARCH AND DEVELOPMENT STUDIES AT ICE

Multiple measures in evaluating the usability of vehicles

An example of this multi-measure approach is a project carried out at ICE in 1991 for the UK Department of Transport. The aim of the project was to develop a simple and practical way of measuring low to medium levels of stress amongst car drivers while the car was being driven. The concern was that a proliferation of driver information systems both inside and outside the vehicle might cause driver stress, with consequential effects on driver performance and road safety. Looked at in the context of this chapter, the concept of stress was being used as a measure of the usability of the information systems.

A review of the literature, together with consultations with other scientists, led us to conclude that the 'true' index of stress was the level of catecholamines in the blood. It is impractical to take such a measure during actual driving. We therefore sought a more feasible method which could be used in the field. Based on our initial investigations we organized a battery of measures, each of which had the potential to correlate with the 'true' biochemical measure. The battery was divided into three broad categories: physiological measures (for example, blood adrenaline, heart rate, respiratory rate); behavioural measures (for example, grip on the steering-wheel, blink rate, body posture); and psychological measures (for example the Stress/ Arousal Checklist (SAC) and the Visual Analogue Scales (VAS)). The complete battery of measures was used in a laboratory simulation of a driving task with different levels of stress, using a panel of subjects over a two-week period. Whilst it is unusual, or unfashionable, to report a null experimental result, we have to say that we could identify no reliable association between any of these measures and the 'true' measure. Grip on the steering-wheel did differ consistently between trials at different levels of stress but this was not statistically significant. It was also counter-intuitive that as stress increased, grip on the wheel decreased!

Despite this disappointing result we proceeded to organize a second series of trials in which a different sample of subjects drove a car on the public roads. A reduced battery of measures was used. Once again the consistency between physiological, behavioural and psychological measures was less than satisfactory at the comparatively low levels of stress in which we were interested. Nonetheless there were some signs that a simple subjective measure (SACL) and the physiological measure of heart rate (average over the duration of the trial, plus the coefficient of variation) could, under the right conditions, prove worthy of further investigation.

Human effort in domestic tasks

We have also attempted to examine the relationship between physical, physiological and psychological measures in the course of a series of studies of human effort in domestic cleaning tasks. The cleaning of surfaces such as kitchen worktops and sanitary ware is a routine domestic activity and can involve a degree of muscular effort, sometimes in awkward body postures. It is to the benefit of the consumer to minimize the work required, by reducing the necessary muscular effort and/or the time taken to complete the task. Changing the formulation of a cleaning product or using a different applicator can change the ease with which it can be applied, used and removed from the surface. The effort required to do this is thus a measure of the usability of the cleaning product.

Typically, these products are evaluated subjectively using a large panel of volunteers. The large panel is necessary because of the inherent variability of subjective measures. This makes evaluation a lengthy and expensive exercise. Our aim was to see whether a direct measure of effort could be derived, which would be less costly to administer and which would yield results that correlate with the results of subjective measures.

We achieved the means to do this by arranging trials in which up to three measures of effort were taken in a variety of tasks and in a range of conditions. We measured the force actually applied to the surface: this required some sophisticated instrumentation which would not interfere with normal use of the surface. Physiological measures – mainly heart rate – were taken continuously by means of a small, portable instrument worn by the subject. Psychological, or subjective, measures of effort were made using Fleishman's (1978) adaptation of Borg's (1973) scale of perceived effort.

These trials have generated some interesting results and have indicated some areas where additional work may be warranted. The results have been the clearest when subjects were dealing with simple, uniform and plane surfaces. In this instance, objective measures have been shown to yield more accurate and consistent results than subjective data, and have been able to discriminate more effectively than subjective measures between people and between activities. In dealing with more complex surfaces the results have been less explicit, due in part to practical issues. For instance, some physiological measures are as sensitive to the context of the experiment itself as they are to differences between tasks or products. In addition, some subjects experienced difficulty in adhering to the (perhaps complex) experimental procedures required in the trials. This tended to introduce variation and anomalous values into the data. It is clear that close control is needed over the experimental protocols if such variation is to be minimized. We have found that on a simple energy expenditure task it is possible to achieve good agreement between the work done, heart rate during the task, and perceived effort. However, in the subsequent cleaning task, the subject's perception of effort is not a very reliable reflection of the work actually done, nor of the heart rate during the cleaning task. Reasonable results can be obtained but only after the subject has been carefully trained in the use of the scale of perceived effort. Heart rate itself is a better predictor of work done ($r = 0.7$) in the tasks we studied than is perceived effort ($r = 0.3$).

IMPLICATIONS FOR THE FUTURE

The studies we have carried out indicate that multiple measure of human activity, at different reductive levels, can generate a rich picture of what is meant by usability, and can add weight to a conclusion which cannot be furnished by the results of one method alone. In addition, such an approach helps to clarify our definition of usability and highlights the fact that a given measure may be only indirectly, and sometimes loosely, related to what we may think of as a 'true' measure. In many of the studies we have undertaken, only an imprecise relationship has been found between what people perceive and what they actually do.

Subjective measures of usability are attractive, as they are normally easy to administer and require little or no special equipment. In the human activities we have studied however, subjective measures appear to be very variable and of doubtful validity. To overcome at least a part of this disadvantage it is necessary to use large numbers of experimental volunteers, and/or to train them carefully in the use of a subjective scale so that it is firmly anchored to the subject's personal experience of different levels of work done. Repeated measures of perceived effort also help to control variability. Such procedures and precautions are time consuming and expensive and even then, the question remains as to how valid the results will be. In many product evaluations it is quite legitimate to argue that the user's perceptions of the experience of using the product are very important for the purposes of marketing, advertising, and initial product use. However, one must also question whether such perceptions are an adequate predictor of the long-term experience of the user with the product; and therefore whether other, perhaps more direct, measures should be used in a complementary fashion to give a more accurate overall conclusion. Provided that some of the practical problems we have encountered can be resolved satisfactorily – and there are no obvious reasons why they cannot – then we see a useful future for the multi-measure approach.

We have some evidence as to the utility of the approach in the eyes of the industrial client. Our studies of domestic cleaning tasks were commercially driven from the outset and were aimed specifically at developing a cost-effective and practical evaluation tool. Despite the methodological problems we have described, we were pleased with the outcome. The organization for whom the work was done has adapted the results of our studies with enthusiasm, and has produced a number of examples of the equipment we developed for its research and development sites in a number of countries. We believe that in addition to offering a cost-effective approach to usability evaluation our work has encouraged this manufacturer at least to examine the concept of usability closely, rather than taking a perhaps more traditional, and superficial, view. At the same time, however, we readily acknowledge the need not only for further development of the method, but for its dissemination and implementation in the evaluation of a much wider range of products.

Our greatest experience in the use of the approach has been where an obvious physical and measurable dimension underlies the concept of usability and which can range quite widely. It is possible that our studies of driver performance were less than satisfactory because the actual level of stress in the task did not span such a range. After all, if two things are to covary, then by definition they must both vary. Nevertheless it would be interesting, and potentially useful, to identify and explore simultaneous behavioural, physiological and psychological evaluation methods in a range of products. For example, we have already mentioned the possibility of using electromyography to evaluate computer input devices. Blink rate may be a potentially useful measure of software usability. The number of postural changes per unit time might indicate whether a seat is comfortable or not. In all such instances, however, it will be necessary first to undertake the basic research to establish the nature and strength of the association between the measures in question.

REFERENCES

BORG, G.A.V. (1973) Perceived exertion: A note on history and methods, *Medicine and Science in Sports*, **5**(2): 90–93.

FLEISHMAN, A.E. and HOGAN, J.C. (1978) *Taxonomic method for assessing the physical requirements of jobs: The physical abilities analysis approach*, Technical Report 3012/R7-8-6, Bethesda MD: Advanced Research Resources Organisation.

PART TWO

Selecting Evaluation Methods

Factors affecting the selection of methods and techniques prior to conducting a usability evaluation

NEVILLE STANTON

Department of Psychology, University of Southampton, Southampton, UK

CHRIS BABER

School of Manufacturing and Mechanical Engineering, University of Birmingham, Birmingham, UK

INTRODUCTION

The idea of a *usability evaluation* is gaining momentum, and the term *usability* is becoming common parlance in product design. This is a welcome shift in emphasis towards 'ease-of-use' in product development. In many respects, the fundamental tenet of usability is that a product should be easy to use. This heightening of interest does not mean that usability (or user friendly, or ergonomically designed, or user-centred design, or consumer-oriented product development) is a new concept. Ergonomists have been beating this particular drum for the past 50 years or so. Less than 20 years ago Ivergard (1976) pointed out that very little research had been published on consumer ergonomics. Although that picture has changed somewhat (there is a growing literature on physical aspects of product use), there is still little published information on cognitive aspects of product use (Baber and Stanton, 1995).

While one can point to the consequences of not considering usability, there is much debate as to what the term actually means. One of the main problems with the term 'usability' is that it means different things to different people. Some suggest that usability is simply another attempt to introduce 'user friendliness' back into product design jargon: usability is simply new wine in old bottles. Others argue that the issues surrounding usability have already been dealt with in 'user-centred design'. Baber (1993) points out, using the analogy of the soupstone, that the term 'usability' takes on individual meaning to each person involved in evaluation to describe whatever they are doing: the individual adds his or her own ingredients. The trouble with this approach is how can we determine if one product is better than another, or indeed if the product has achieved some acceptable benchmark? Clearly this matter needs to be resolved as a matter of urgency, particularly in the light of recent legislation which makes usability a legal requirement in some products! This is a rather ridiculous situation given the debate and controversy surrounding the concept of usability (Stanton and Baber, 1992).

WHAT IS USABILITY?

While it is possible to indicate the necessity for usability in product development, as a concept it has proved remarkably resilient to definition; we all know what it is, but have difficulty reaching an agreed, coherent definition (it is our own personal soupstone) which will allow recommendations to be made concerning how best to make something more 'usable'. This is the first, and perhaps most important, stumbling block in determining methods appropriate to evaluation. If we cannot agree on what usability is, how can be hope to measure it? It is likely that different definitions of the concept will lead people to measure different aspects of product use. This suggests that a usability evaluation may not have a common standard between individuals. If usability is to be more than an ephemeral concept, we must agree on its constituent ingredients. Stanton and Baber (1992) draw upon a decade of work represented by Shackel (1981), Eason (1984) and Booth (1989) to suggest the following factors which serve to shape the concept of usability and define its scope.

1 *Learnability*: A system should allow users to reach acceptable performance levels within a specified time.
2 *Effectiveness*: Acceptable performance should be achieved by a defined proportion of the user population, over a specified range of tasks and in a specified range of environments.
3 *Attitude*: Acceptable performance should be achieved within acceptable human costs, in terms of fatigue, stress, frustration, discomfort and satisfaction.
4 *Flexibility*: The product should be able to deal with a range of tasks beyond those first specified.
5 *The perceived usefulness or utility of the product*: Eason (1984) has argued that '... the major indicator of usability is whether a [product] is used ...'. As Booth (1989) points out, it may be possible to design a product which rates high on the LEAF precepts (1–4 above), but which is simply not used.
6 *Task match*: In addition to the LEAF precepts set out above, a 'usable' product should exhibit an acceptable match between the functions provided by the system and the needs and requirements of the user.
7 *Task characteristics*: The frequency with which a task can be performed and the degree to which the task can be modified, e.g. in terms of variability of information requirements.

8   *User characteristics*: Another section which should be included in a definition of usability concerns the knowledge, skills and motivation of the user population.

We may argue over the relative merits of different ingredients and the labels we give them, but this rarely becomes more than a exercise in semantics. ISO 9241 goes some way towards incorporating the above factors, but we feel that it falls short of a comprehensive definition in an important way. From reading ISO 9241 (at the time of writing this was still unreleased), we feel that usability has been defined by what can be measured: usability is what usability evaluations do. This appears to largely concentrate on the LEAF precepts listed above (learnability, effectiveness, attitude and flexibility). We believe that haste in producing the definition of usability should be chastened by rather more circumspect consideration about what is meant by usability.

TECHNIQUES FOR EXAMINING USABILITY

The reader will not be surprised to learn that each of the various factors that make up usability has spawned particular approaches to usability evaluation. In this section we present approaches related to aspects of product development. We are particularly concerned that reliance upon one approach exclusively or a very narrow definition of usability, could lead an individual to perform a limited usability evaluation. This concern should become obvious when we map evaluation methods onto the ingredients of usability.

Usability in the design process

Traditionally, usability assessments have been performed at the end of the design cycle, when a finished product can be evaluated. However, it has been noted that the resulting changes proposed may be substantial and costly. This has led to a call for usability to participate in earlier aspects of the design cycle. The most obvious manner to collect information about how people perform a task is to watch them do it, or ask them about how they do it.

- *Observation* of human activities may afford the collection of data about the real interaction between the human and the machine. Yet what is observable might not tell you for example, about the decisions being made, or the alternatives not selected.
- *Interviews*, on the other hand, may provide a quick way to get to the unobservable decision-making process, particularly if the interviewee is describing a recent event, and how they dealt with it. However, it has been found that when people describe events, the description differs from what they were actually doing at the time. A solution to this is to get them to 'think aloud'

(concurrent verbal protocol), but people may actually change the way they do things to make them easier to describe!

This information may then be analyzed by a variety of means, such as task analysis, link analysis, time line analysis and layout analysis; these are described below:

- *Task analysis* has many derivatives, British ergonomists use hierarchical task analysis (HTA; see Stammers, Chapter 23). This allows us to describe human behaviour in units from a hierarchical presentation, describing behaviour in terms of goals, plans and actions. However, HTA often only presents ideal behaviour, or an agreed consensus of ideal behaviour. The task analysis may then feed into user requirements and system specification. However, this mapping assumes that existing task practices are ideal. It may be necessary to further abstract task goals at a general level before proposing desired human activity and specifying system requirements.

- *Link analysis* (Drury, 1990) allows us to examine the way humans use displays, and to catalogue the frequency with which they switch between displays and the time spent at each display. Together with an expert rating of the importance of each display, this allows us to present an optimum layout of the displays.

- *Time line analysis* (Miester, 1989) enables the study of the task sequence and the operations performed. System bottlenecks may become apparent, which prompts the need for redesign. A new design may then be rerun and reanalyzed to see if the bottleneck in the system is now cleared.

- *Layout analysis* (Easterby, 1984) is a means of examining display and control layouts using four criteria: functional classification, importance of item, sequence of use, and frequency of use. Based on this analysis the layout may then be subjected to improvement and redesign.

The criteria for acceptance of methods will include the time limit of the project, resources available, skills of the practitioners, and the stage of design. For example, a technique such as link analysis is relatively easy to perform and does not require knowledge of ergonomics in its use. HTA, however, requires considerable experience if it is to be used effectively.

Usability evaluation and assessment may enter all stages of the design process, from a requirement analysis, through initial design specification and prototyping, up to and including the working product. Consider the four classes of evaluation method proposed by Whitefield *et al.* (1991): analytical methods (e.g. TAFEI), specialist reports (e.g. layout analysis), user reports (e.g. interviews) and observational methods (e.g. HTA and link analysis). Each method is appropriate to certain stages of the product development cycle. This is illustrated in Figure 5.1.

As Figure 5.1 suggests, analytical methods are mainly appropriate prior to product development. At this point there is a good opportunity to iron out many potential problems, at relatively low cost. During product design specialist reports provide the main input. After product development most of the methods appear to be used. However, this is the most costly point in the design process to make

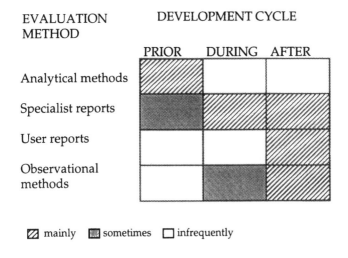

EVALUATION   DEVELOPMENT CYCLE
METHOD

Figure 5.1 Evaluation methods for the design process

changes and therefore underlines the impetus to include usability considerations earlier on. In this way, we may view usability input as a coarse-to-fine design approach. Early on in the design process it is able to offer a range of alternatives, which are successively refined until the product is implemented. This refinement goes hand in hand with the prototyping and evaluation process. Although Figure 5.1 may represent our notion of current of practice, this does not necessarily mean that it is a desirable position. We are particularly keen to involve user evaluations earlier on in the design process.

Rapid prototyping is one means of achieving this. Developing prototypes enables the users to contribute to the design process. In this way identified problems may be reduced well before the final implementation of the product. The contributions of the user do not only take the form of expressed likes and dislikes, but may be more objective in terms of performance testing. This participation may also provide the designers with a greater insight into the needs of the users. This can be particularly important when the users may not be able to clearly express what features they desire of the system, as they are likely to be unaware of all the possibilities that could be made available. However, the participants in prototyping ought to be either 'real' end users or, at least, representative users. Organizations often employ 'product specialists' to act as users, but these specialists will develop expertise which differs from that of 'real' users. It would make more sense to use an *ad hoc* panel of users in absence of access to the 'real' users.

In the rapid prototyping process, successive generations of the product are evaluated and the requirement specification finely tuned until the final product emerges. Rapid prototyping is a means of introducing the typically 'late' evaluation techniques (illustrated in the 'after' section of Figure 5.1; i.e. human factors methods used after the developmental cycle is complete) earlier in the development cycle. Thus changes are likely to be more acceptable in terms of cost and therefore more likely to be implemented. Rapid prototyping recognizes that product development ought to be incremental, and allows usability to be incorporated into a succession of stages. As such it manages to infiltrate the design process in a subliminal way, rather than being a cumbersome add-on. Some organizations conduct prototyping trials in 'usability labs'. This formalizes the evaluation and provides the opportunity to capture behaviour from a variety of sources simultaneously, e.g. keystrokes, eye movements, verbal protocols and physical activity. These data are recorded by the computer, eye cameras, audio tape recorders, and video recorders. This enables the analysis to be conducted at the analyst's leisure, and the users' behaviour can be scrutinized in fine detail. However, the mass of data generated will not cover the fact that much of it will be meaningless in the absence of performance criteria.

Evaluating existing products

Many writers have developed checklists for the evaluation of products in terms of their usability. One of the most detailed and useful was developed by Ravden and Johnson (1989). Checklists will be most useful for factors 2, 3, 6 and 7 (effectiveness, attitude, task match and task characteristics) listed above. They can be used to elicit the user's attitudes, or to evaluate the relationship between task performance and the computer. However, checklists are not the best way to assess items 1, 4, 5 and 8 (learnability, flexibility, perceived usefulness and user characteristics). These could be considered either through experimentation (for 1 and 2, learnability and effectiveness) and observations (for 1 and 5, learnability and perceived usefulness), or through detailed specification of performance objectives (4, flexibility), or through development and testing of scientific theories (8, user characteristics). It is these approaches which are currently proving difficult to use. The main problem lies in the question of how to define performance criteria and experimental measures. This is mainly due to the fact that performance will be highly context dependent. It will depend upon the nature of the task being performed and the knowledge that users bring with them to the interaction. In the absence of coherent theory, one cannot model user characteristics. Therefore we are left with an engineering, rather than scientific approach. This means that the most promising way to effectively measure usability is by using real users of the computer system in question, for example, when prototyping in system development. Many of the methodologies are similarly heuristic in nature, and are neither proven nor validated. This means that they should be used with caution. Other approaches are available to researchers, e.g. verbal protocol, experimentation, task analysis, simulation. They can be very useful in product evaluation, but need to be performed after training in order to yield meaningful data.

Evaluating early concepts

Existing products can be evaluated quite simply by asking people to use them and then using a range of techniques to observe and analyze this usage, but it is a far harder proposition to evaluate conceptual products, i.e. paper-based designs. Yet it is while the product is in its conceptual stage that the designer will have the most opportunity to incorporate usability into the product's design. A problem with usability is that, while it is possible to base specifications on products with which the designers are familiar, these products may not in themselves be 'usable'. As these familiar products are altered through redesign, then the usability specifications will necessarily alter; change the product and you will change the nature of the product's use.

A number of packages exist which allow designers to prototype proposed mock-ups of products. While such prototypes can be used to apply specific guidelines of product design, they cannot be used in an evaluation of usability unless they form part of a rapid prototyping schedule. Furthermore, it is only possible to prototype products when the task has been described adequately. This point is illustrated by Carroll and Rosson's (1991) notion of a task–artefact cycle. Basically, the design of a product will influence its use, which will influence the user's goals, which will influence the design of a product. They suggest scenario analysis or storyboarding as a means of capturing this cycle. This allows designers to maintain a flexible attitude to what they are designing and for whom they are designing. The designer's conception of how a task is performed is very different from that of a user. This means that a gulf may well exist between design and user requirements (Norman, 1988). User requirements capture techniques have been developed and a number of techniques which exist in ergonomics can be used successfully for this purpose, e.g. task analysis (Diaper, 1989). This can then provide objective information concerning real task requirements, which, in turn, can form the basis of specifications. However, such specifications will only provide 'static' product data, i.e. information concerning how the product ought to look, not necessarily how it ought to be used 'in anger'. Stanton and Baber (Chapter 24) present a technique (TAFEI) which they claim can be used to analyze human interaction with products while development is still at the conceptual stage. This technique is potentially very useful to product designers.

SELECTION OF METHODS AND TECHNIQUES

To assist in the process of selecting methods and techniques to be employed in the usability evaluation Stanton and Baber (1994) devised the heuristic chart shown in Figure 5.2. This chart could enable product designers to determine which method(s) is/are appropriate for their usability evaluations. The list of methods is not intended to be exhaustive, and we invite readers to add their own. We are trying to highlight the need to consider the demands and constraints being placed on the project which are likely to affect the appropriateness of evaluation methods.

We have reduced the number of questions that the designer needs to ask the design team and project. The purpose of these questions is to make the decision of which method to choose based upon a considered judgement, rather than an *ad hoc* selection. We are not necessarily proposing that anyone stick rigidly to the

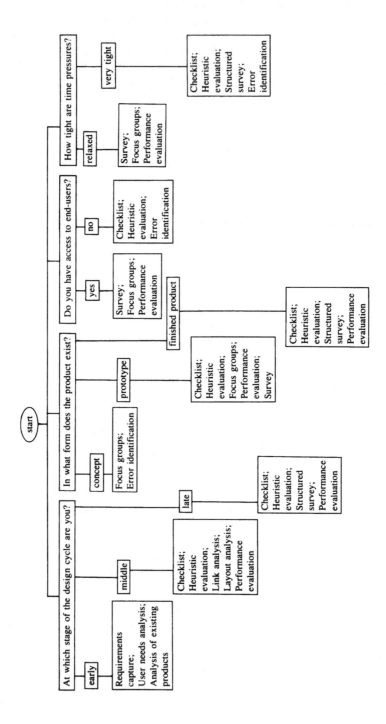

Figure 5.2   Selection of methods and techniques

procedure; rather that they think hard about why they have selected one method in preference to another. We feel that the selection process should be the subject of further research and that it is desirable that the process should be as objective as is possible.

In brief, our approach makes explicit the questions that designers would ask before opting for an evaluation method. The factors to be considered are:

- the stage of the design cycle
- the form of the product
- access to end users
- time pressures.

First, the stage that the *design cycle* is in needs to be ascertained, be it early (whilst the product is still in the conceptual stage), middle (when a prototype of the product has been produced) or late (when the final version of product has been produced). Methods are appropriate at different points in the design process as we have indicated. The second question asks if the *product* (or a similar product has to be *in existence* before the method can be used. Some methods do not rely upon product existence. The third question relates to the *availability of end users*. Again, some methods can be used independent of access to end users. The fourth question determines the *time window* of the project since some methods can be performed relatively quickly. At present checklists (see Johnson, Chapter 20) appear to be most popular in current practice (Mirza and Baber, 1994). Consideration of Figure 5.2 makes it clear why this is the case. The use of a checklist in the hands of an experienced designer and ergonomist can make it a very cost-effective evaluation method. Examples of other methods can be found throughout this volume.

CONCLUSIONS

In order to be used, and used effectively, any product should be 'usable'. This is not simply a truism (although at first glance it may appear so). Products should be designed to conform to basic principles of usability. This chapter has sought to outline some of these principles, and to relate them to methods for usability evaluations. We would caution sole reliance upon outcome measures in usability evaluations, as they do not necessarily guide redesign of products. In usability evaluations we believe that product *use* is of greatest importance, and therefore feel that process measures are of most benefit. This observation is made on the premise that an understanding of how the product is being used will ultimately provide more useful information to the designer than telling him or her that the product was not usable. Usability can facilitate the design of human-centred products which are not simply 'user-friendly' but also useful. Our interpretation of the concept of usability suggests that it is important to ensure a good task match between products and users. This task-oriented, rather than technology-oriented approach is indicated by the methods we propose.

REFERENCES

BABER, C. (1993) Usability is useless, *Interfaces*, **21**: 24–25.

BABER, C. and STANTON, N.A. (1995) Task analysis for error identification: A methodology for designing error tolerant products, *Ergonomics*, **37**(11): 1923–1941.

BOOTH, P. (1989) *An Introduction to Human–Computer Interaction*, London: LEA.

CARROLL, J.M. and ROSSEN, M.B. (1991) Deliberated evolution: Stalking the view matcher in design space, *Human–Computer Interaction*, **VI**, 281–318.

DIAPER, D. (1989) *Task Analysis for Human–Computer Interaction*, Chichester: Ellis Horwood.

DRURY, C.G. (1990) Methods for direct observation of human performance, in: J.R. Wilson and E.N. Corlett (Eds), *Evaluation of Human Work*, London: Taylor & Francis.

EASON, K.D. (1984) Towards the experimental study of usability, *Behaviour and Information Technology*, **3**(2): 133–143.

EASTERBY, R. (1984) Tasks, processes and display design, in: R. Easterby and H. Zwaga (Eds), *Information Design*, London: Taylor & Francis.

IVERGARD, T.B.K. (1976) Ergonomics and the consumer, *Ergonomics*, **19**(3): 321–329.

MIESTER, D. (1989) *Conceptual Aspects of Human Factors*, Baltimore, MD: Johns Hopkins University Press.

MIRZA, M. and BABER, C. (1994) The role of ergonomics in the testing and evaluation of consumer products, presented at EUROLAB 1994, April, Milan.

NORMAN, D.A. (1988) *The Psychology of Everyday Things*, New York: Basic Books.

RAVDEN, S. and JOHNSON, G. (1989) *Evaluating Usability of Human–Computer Interfaces: A Practical Method*, Chichester: Ellis Horwood.

SHACKEL, B. (1981) The concept of usability, *Proc. IBM Software and Information Usability Symposium*, pp. 1–30, New York: IBM.

STANTON, N.A. and BABER, C. (1992) Usability and EC Directive 90/270, *Displays*, **13**(3): 151–160.

STANTON, N.A. and BABER, C. (1994) *A Pragmatic Approach to the Design and Evaluation of User Interfaces*, Tutorial Notes for The Ergonomics Society Annual Conference, University of Warwick, 19–22 April 1994.

WHITEFIELD, A., WILSON, F. and DOWELL, J. (1991) A framework for human factors evaluation, *Behaviour and Information Technology*, **10**(1): 65–80.

CHAPTER SIX

Information ecologies, experience and ergonomics

DICK RIJKEN and BERT MULDER

Utrecht School of the Arts, 1200 CL Hilversum, The Netherlands

INTRODUCTION: INTERACTIVITY AND USABILITY

The Interaction Design department at the Utrecht School of the Arts offers a four-year curriculum in Interaction Design. Connected to the course is an expertise centre that links the curriculum to the outside world. The Centre for Interaction Design offers education, contract research, and consultancy to industry.

We view Interaction Design as a human-centred design discipline that builds on traditional disciplines (various design disciplines, social sciences, technology, philosophy) combined with its own body of knowledge and skills. Usability evaluation is a crucial activity in the design process, but different types of applications often pose different usability problems. Computer applications that support tasks in work environments must be easy to use, while computer games can be fun exactly because of the challenges they pose. New developments in information technology (graphical user interfaces, interactive multimedia, computer networks) change the nature of interactivity: active users engage in psychologically rich interactions with complex information environments. As a discipline that aims to optimize the relationships between people and artefacts, ergonomics needs to address these new and more complex forms of interaction.

The quality of interaction depends on how well the goals of the interaction are satisfied in their context during the process of interaction. The cognitive relationship between a manager at work and a decision support system and the emotional relationship between a teenager at home and a computer game are characterized by different sets of goals. We argue that this requires a more general framework for thinking about (complex) interactive systems and (complex) interactions. Our approach to these issues is outlined below.

INFORMATION ECOLOGIES: COMPLEX WEBS OF RELATIONSHIPS

In this section we argue that we need to review our thinking about interactive systems: *what* is the user interacting with?

We are moving from the simple single-user, single-product situation to computer-supported collaborative work, the design of interfaces for entire organizations and worldwide information environments. Everything will be connected to everything else. When the environment of an interactive system includes multiple platforms, multiple applications, global networking capabilities and so on, it is no longer clear what or where the boundaries of the system are, and who or what the user is interacting with. An information system simply becomes infinite from a user's point of view.

An ecological perspective

Ecology is the study of systems or organisms in relation to their environment. We propose an *ecological perspective* on interactive systems where an individual user not only *interacts with* a system, but rather *acts inside* an information ecology. In complex ecological systems, the relationships between species form an intricate web. In information ecologies, users, systems and hybrid forms define and influence each other reciprocally. They are significant components of each other's environments – shared information spaces structure their interactions. The Internet is a good example of a large information ecology. Rheingold (1993) discusses how people in social contact on the Internet have established communities where rules of conduct have evolved in self-regulating ways.

The ecological perspective stimulates us to ask new kinds of questions that can aid our thinking about design and ergonomics: what are the structural and temporal characteristics of information ecologies (from a user's point of view), how do they affect the quality of interaction? How can ergonomics as a discipline deal with these new questions?

Some characteristics of information ecologies

In information ecologies, individual products and individual users are always elements of larger wholes. The patterns of interaction between users and systems cannot be controlled or predicted beforehand. Users in information ecologies can function adequately even though the information may be inconsistent and delayed, and decisions can be made based on imperfect knowledge. The (implicit) order unfolds in the course of action and is inherently unpredictable. In terms of structural characteristics, the ecological perspective probes us to distinguish between implicit and explicit order.

This also brings us to temporal issues like adaptation, growth and evolution.

Users shape and are shaped in information ecologies. Over time, users and systems continuously adapt as they learn about and modify their environments. The development of many future systems will have an evolutionary character. Networks of databases will grow and evolve according to how, by whom, and how often they are used. Radical change is improbable and small additions can have large-scale effects.

Design and ergonomics for information ecologies

How can ergonomics as a discipline address these structural and temporal issues? In information ecologies, we cannot *design* new situations, but must allow them to grow and evolve. Designers facilitate processes of growth in existing ecologies and pre-facilitate it in the creation of new ecologies. If we see ergonomics as a form of reflection, it must provide new methods and techniques for continuous reflection on information ecologies. An individual design in a single-user, single-product context generally has a set of well defined goals for the interaction, that can be evaluated. In terms of structural aspects of ecologies, we must no longer only ask what a specific system is for, but also develop and apply quality criteria for entire ecologies of systems. We must identify attributes of information ecologies that are relevant for usability evaluation and find ways of measuring them quantitatively or qualitatively.

In terms of temporal aspects, ergonomics must find ways of evaluating adaptation, growth and evolution of the symbiosis between user and environment. Developments over time may dramatically influence the characteristics of system, user, and their patterns of interaction. Continuous usability evaluation becomes a necessary strategy. Since it is not known beforehand what may be relevant to monitor, the collected data must be broad enough to allow multiple frameworks for interpretation.

What communities does a user belong to and what are the relationships with other members? Can we measure the adaptive quality of systems? How well are products suited for other uses? Or for other users? To what degree does the quality of an individual system depend on specific circumstances? How does any individual system relate to other systems in its vicinity? How will the adaptation of users affect their functioning elsewhere? What growth paths can be expected and anticipated? These are new kinds of questions that can be asked in evaluation studies.

An example

In one instance we offered users the possibility to provide comments (in a free text format) on an administrative system. They were told that any kind of feedback was appreciated and would be used in efforts to improve the quality of the system. They could browse through their own comments as well as through those of other users at any time. When their messages were read, users were notified, and when their suggestions were used, they received a report of how their comments related to changes in new versions of the software. Over time, the usefulness of their suggestions improved as they learned what kinds of comments caused what kinds of improvements. This experiment resulted in an enhanced sense of involvement and responsibility on their part.

An approach such as this illustrates a new role of ergonomics and how it necessitates a seamless integration with design.

EXPERIENCE: THE CONTENT OF INTERACTION BECOMES MORE COMPLEX

We also need to review our thinking about interactivity: what is the *content* of interaction? Many interactive systems can still be viewed as *tools* to support more or less cognitive tasks (word processors, CAD systems, drawing programs, etc.) or *media* with some 'message' to be communicated (CD-I, Point-Of-Information, Point-of-Sale, etc.). But this is changing. The medium of television is becoming a shopping tool and word processing tools transform into hypermedia environments. Media users become active navigators, while tool users lose control or initiative when interacting or communicating with other people or with software agents. Computer games can feature immersive environments with immediate and intuitive loops of control and feedback.

The relationships between users and systems have become more complex. Many new kinds of products are not fully understandable in terms of information and communication. Moving from simple text and graphics to interactive multimedia and virtual reality, there is a shift from information to experience. Future information environments will feature rich and powerful experiences that will be multisensory and multifaceted (physical, cognitive, emotional, motivational, even spiritual).

Experience in information ecologies

Regarding experience in complex information ecologies, the challenge for ergonomics as a discipline is twofold. We need to broaden as well as deepen our thinking about experience:

- experience can be very rich, involving the body, the mind, and even the soul;
- experience is very much an inner process that is hardly subject to behavioural observation.

In the following we show how these two questions can be integrated and present some tools and techniques that we use in practice: the 'Utrecht matrix' and phenomenological analysis. The 'Utrecht matrix' is a simple tool for dealing with the richness of experiences, and the phenomenological analysis is applied to attempt to deal with the subjective wholeness of experiences.

The richness of experience: the Utrecht matrix

In order to address the potential richness of experience, we need to broaden our view of users. People are not just information processors while using information technology. Below we outline some of the issues that ergonomics needs to address:

- *The body*: There is a wealth of information on physical ergonomics. But what is happening to our bodies? Where is my body when I am telepresent? What is my body in virtual reality?
- *The mind* (cognition, emotion, the will): Cognitive ergonomics has yielded much knowledge about cognitive aspects of interactivity. But what about our feelings, our irrational tastes. What about the will? How can we make users reflect on their actions, care about the interaction and motivate them to invest time and effort?
- *The soul*: Increasingly, people are being confronted with choices that are moral or ethical in nature when using interactive systems. An understanding of these issues will be expected of designers and evaluators.

We use a simple, but very effective tool that helps us to identify possible relationships between users and systems: the 'Utrecht matrix' (see Figure 6.1). It features very general models of people and systems: every person has a body, a mind and a soul; every interactive system has content (information, functionality accessible to users), structure (the apparent organization of content), behaviour (the system's manifest behaviour and the user's evoked behaviour), and appearance (physical, perceptible form).

The Utrecht matrix can structure our thinking about usability. It functions as a checklist during analysis, design and evaluation activities as it stimulates us to identify the critical cells (given the goals of the different stakeholders) during

system: user:	appearance	behaviour	structure	content
body				
mind / cognition				
mind / emotion				
mind / the will				
soul				

Figure 6.1   The Utrecht matrix.

analysis, design and evaluation activities. The matrix probes us to quickly examine many relationships between users and systems that may need to be evaluated.

Using the matrix involves the following steps:

- Identify the goals of different stakeholders (developers, buyers, users) and identify critical cells (critical success factors) that relate to these goals (analysis).
- Prioritize the cells with regard to design goals and 'aim' the design at the content of the most important cells (design).
- Find tests to evaluate the crucial cells (evaluation).

Take, for example, an 'edutainment' application for children in the form of a game that requires topographical knowledge to complete. The goal of the parents (the reason why they buy the game!) is that the game educates the child. The goal of the child is to have fun playing. The goal 'inside' the game is try to find a hidden treasure. Critical cells are:

[content – cognition]	The child must learn about topography.
[content – the will]	The prospect of finding the treasure must motivate the child.
[behaviour – emotion]	Navigating through the game's 'spaces' should be fun to do.
[appearance – emotion]	The game should look attractive.

For each of these cells, we can attempt to devise tests that will tell us whether we have succeeded in satisfying these goals. The Utrecht matrix can make trade-offs visible: 'navigation is smooth (and therefore pleasurable) if we use low-quality graphics, but that makes the appearance less attractive. . .'. What is more important? Are there alternative solutions?'

When using the Utrecht matrix for usability evaluation, the problem is that there are many 'blank spots' in the matrix – for some cells there are few methods and techniques available for evaluation purposes! Whereas physical and cognitive ergonomics have provided a wealth of methods, tools, and techniques, the more qualitative issues (emotion, the will, the soul) can only be tested with very general techniques like observation studies, interviews and questionnaires. There is work to be done: a lot of borrowing from different social sciences, and a lot of fresh thinking for ergonomics as a discipline. In the following section we present one of our attempts to deal with these kinds of issues.

The wholeness of experience: phenomenology

Many existing usability evaluation methods, techniques and guidelines have their origins in the field of office automation. They come from a context of office workers and work responsibilities with a strong cognitive bias. Also, many of them focus on collecting observable data as a starting point. But how can we evaluate the quality of interaction in the context of rich experiences? Is there such a thing as emotional ergonomics?

Usability evaluation must place more emphasis on the situated nature of the interaction, from an 'external' as well as from an 'internal' point of view: a description of the environment of use as well as an understanding of the user's subjective experience of the situation. Traditional analytical approaches require a new and more careful application, to avoid missing qualities of the inherent wholeness of the user's experience. We must find methods and techniques for dealing with inner experiences as well as with observable behaviours. In terms of ecological validity, we must not only make descriptions of the context of use and of the tasks at hand, but also try to really *understand* experience in a way that does justice to the inner richness involved.

The inherent wholeness of the user's experience is hardly subject to objective measurement by any observation methods or techniques at all. Ergonomics needs to find valid ways of dealing with these inner processes without falling into the pitfalls of gratuitous introspection and subjectivism.

We are exploring the possibilities of structured analysis of experience as used in phenomenological psychology (Varela *et al.*, 1991; Winograd and Flores, 1986). The tradition of existential phenomenology has dealt with the problem of formulating a psychology of experience from the actor's point of view. Phenomenology views people as shapers of the world that shapes them. All action is situated; it is always embedded in a multifaceted context (socially, physically, etc.). Consciousness is always intentional; people are always conscious of *something*. People interpret their situation subjectively and act upon their perceived meaning – experience and action are two sides of the same coin.

Preliminary 'testing': phenomenological (task) analysis

Most task analysis methods are very observational in nature. But suppose we want to gain a deeper understanding of a task before we start any design activities. We use phenomenological analysis as a method for capturing the essence of a task (that which is general over people and contexts). What does the task mean in the user's frame of reference? What elements of the task generate meaning? For a phenomenological analysis of some phenomenon (e.g. a task), we actively use variation as a method for collecting many different outlooks on the phenomenon. We can compare various examples of experiences (same task in different situations, with different users, with different tools) and try to discover what is common to them. We can use our own experiences, other people's experiences, possibly including deviations and exceptions, study the phenomenon in various cultures, imagined variations, etc. When we are done, what is common or relatively constant, is inter-subjective and assumed to capture the experiential essence of the phenomenon.

Phenomenology in practice

For example, if we need to evaluate a decision support system, we try to discover the essential attributes of the 'decision-making' experience. Part of the solution lies in the cognitive structure of these types of tasks, involving the weighing of alternatives and other such rational activities (cognitive science can help here), but let's ask a different question: what does it really *mean* to make decisions in the context of individual responsibilities in an organization or some other situation? How does the difference between hierarchical and more loosely structured organizations influence this process? Emotionally, for instance? How does it feel to be responsible? What kind of stress results from these tasks? Phenomenological analysis is an attempt to gain an understanding of subjective qualities of tasks, while remaining independent of specific individuals.

In practice, thorough phenomenological analysis can be time-consuming. We have experimented with shorter, informal methods. Four or five descriptions of experiences in short, one-page scenarios, are less time-consuming, but they do require a more qualitative inspection of experiences. In order to investigate the idea of 'to search', we can imagine and describe the following experiences: an old man is looking for his newspaper; a mother is looking for her child in a shopping mall; a driver tries to find his way through town; a young woman, late for work, looking for her car keys; a student browses through a library, looking for nothing in particular (just looking. . .).

Also, freeform observational studies and interviews can provide relevant information. Generalization then often yields a fair understanding of the more qualitative aspects of user experiences (search tasks in the example) and sharpens the alertness for qualitative aspects of usability in more traditional observation studies. In order to optimize ecological validity, we do usability evaluations 'on the spot', rather than in usability labs. See Rijken (1994) for examples of phenomenology in education.

Phenomenology is a form of reflection on experience that attempts to capture the essences of experiences. We have found that it requires (in the observer) a sensitivity for qualitative aspects of situations as well as an ability to analytically reflect on experiences. So who does phenomenological research? The designer? The evaluator? Again, we see that the distinction between design and ergonomics becomes less distinct and that a seamless integration of the two approaches is required. Usability evaluation has become a true necessity in the development process, now more than ever.

CHALLENGES FOR ERGONOMICS

In summary, we see that the following trends have caused us to examine the necessity of new methods and techniques for usability evaluation:

- Systems have become more complex. Users act inside complex information

INTERACTIVE SYSTEMS		
complexity of system: content of interaction:	single-user-single-product	complex environments
information	physical and cognitive ergonomics	physical and cognitive ergonomics + considering structural and temporal attributes of information ecologies in usability evaluation
experience	physical and cognitive ergonomics + usability matrix, phenomenology	**ecological ergonomics:** **evaluation of rich experiences in complex, dynamic environments**

Figure 6.2 Challenges for ergonomics: interactive systems.

ecologies that evolve over time. This trend has stimulated us to pay attention to complex relationships and find ways of tracking usability over time, using broad collections of data.

- The relationships between users and systems have become more diverse and complex. New and different kinds of interactive systems and products have appeared on the market that must be evaluated (computer games, interactive entertainment). Also, new systems feature interactions that combine information, communication, and multimedia, resulting in rich experiences. This trend has stimulated us to find ways of dealing with more qualitative issues (the Utrecht matrix, phenomenology).

If a design aims at establishing meaningful relationships with users in many respects, then we must know how to find out whether the attempts have been successful. In Figure 6.2, we can see in the lower right corner that everything comes together into what we may call 'ecological ergonomics', a true challenge for ergonomics as a discipline.

REFERENCES

RHEINGOLD, H. (1993) *The Virtual Community: Homesteading on the Electronic Frontier*, Reading, MA: Addison-Wesley.

RIJKEN, G.D. *et al.* (Eds) (1994) Education: Interaction design, *SIGCHI Bulletin*, **26**(3): 70–79.

VARELA, F.J., THOMPSON, E. and ROSCH, E. (1991) *The Embodied Mind: Cognitive Science and Human Experience*, Cambridge, MA: MIT Press.

WINOGRAD, T. and FLORES, F. (1986) *Understanding Computers and Cognition: A New Foundation for Design*, Reading, MA: Addison-Wesley.

CHAPTER SEVEN

Do-it-yourself usability evaluation: guiding software developers to usability

HANS BOTMAN

Utrecht Institute for Higher Education, 3563 AS Utrecht, The Netherlands

INTRODUCTION

Since 1989 I have been working with a large Dutch software company that develops tailor-made systems for medium-size and large companies, with a focus on banking, insurance, professional services and government. The company provides full services from design, development and implementation to maintenance and consultancy. The company flourished in the 1970s and 1980s when large mainframe-based batch systems where developed. Usability was not important since systems were mainly batch (off-line) and were used by small groups of specially trained users. But as both the computer and the terminal were becoming more popular, the group of users diversified and on-line subsystems were added to the batch. Although the company was well aware of these changes the main focus was still on functionality. Testing concentrated on 'bugs' and functionality, and neither of these tests focused explicitly on usability.

As users started to complain about unfriendly interfaces and inconsistencies, a growing awareness arose within the company, leading the way for a new specialism: the human–computer interaction specialists, cognitive psychologists, industrial design engineers and ergonomists who specialized in human–computer interaction, interface design and usability evaluation. For a new actor in an already overcrowded arena, it was difficult to gain acceptance. Access to projects was limited and in many cases restricted to some simple advice and a heuristic evaluation (expert view) of the ready-to-ship product. These evaluations were quite frustrating: too late, with too little time and no budget to implement changes. So, on the one hand, there was a growing demand for usability expertise, but on the other the main demand was for 'knowledge and checklists'; not for expertise. Colleagues mainly asked for advice or some simple rules on how to develop user-friendly software.

Unfortunately, software development is ruled by abstract, logical thinking. Software developers lean heavily on 'checklists'; simple lists with 'golden rules' which tell you what to do in what order. In our view, developing user-friendly systems is not a matter of using the right 'checklist'. Usability requires a different attitude, in which the focus is on the users, and the computer is to be seen as a tool that can help people improve their tasks. Although a good checklist might be helpful to a professional human–computer interaction specialist, the checklist by itself is not enough to change the attitude towards user interface design.

If you can't beat them, make them join you. The logical thing to do was to provide training for developers in human–computer interaction and to teach them some simple methods to improve their usability evaluation capabilities. Show developers that usability engineering can be learned but is in fact a profession and they will call on you when they need professional help. This strategy should both improve the quality of user interfaces and pave the road for a flourishing group of usability specialists. Thus 'do-it-yourself usability evaluation' (DIY) was developed.

DIY is based on 'discount usability engineering' a term introduced by Jakob Nielsen (1990). Using DIY is not a guarantee for usability. It focuses on aspects like guessability and learnability, aimed at finding usability bugs. The most important goal of DIY is to create usability awareness amongst software developers. In fact, if every developer had kept the DIY steps in mind while developing software, many of today's irritating usability bugs might never have occurred.

DO-IT-YOURSELF USABILITY EVALUATION

Many software development projects are characterized by strict budgets, tight schedules and a focus on functionality. In these cases, where usability is not a major focal point, you won't be able to raise hands for a professional usability evaluation. If the budget is too tight to hire professional help, some easy-to-use methods for software developers might be a solution. As Nielsen (1990) argued, user-conscious software developers can achieve reasonably good results by following some simple rules of thumb. The do-it-yourself usability evaluation is at hand if there is no time to use or develop a special method, if it is not possible to involve extensive user-testing, and if there is not enough time for thorough problem/situation orientation.

The DIY usability evaluation should help software developers to increase their 'end-user consciousness'. DIY has two goals: in the short term it should help software developers improve their products: to evaluate and improve the interfaces which they design or build. In the long term these evaluations should change the attitude of developers and thus improve their initial designs, making DIY a natural part of the developers' approach to software development. If developers change their attitudes, many usability errors can be prevented, resulting in improved quality and in lower development costs.

The DIY evaluation contains six steps. It is not necessary to follow all of them; depending on the situation and the amount of money and/or time available, just two or more different steps can be used. The steps are:

1 Orientation on the users and their environment;
2 Expert view;
3 User's view;
4 User testing;
5 Evaluation report;
6 Recommendations for improvement.

(1) Orientation on the users and their environment

Distinguish the different user groups and the environment in which the application will be used. Very important factors are the goals and tasks of the user. Software development is often separated from the user environment. In some cases the user group is unknown, and in others the users are difficult to access. But access to the users is vital for good software development. A good orientation requires visiting the real end users (and not only their managers). Another prerequisite is that the developers listen to the users instead of trying to convince the users. The orientation consists of three parts:

- goals of the organization and users;
- advantages and disadvantages of current jobs, tasks and tools;
- characteristics of users and their environment.

An orientation on the user can change the developers' attitudes. They might become more user-focused once they have seen and heard the users and their everyday problems.

(2) Expert view

Evaluate the application using the usability guidelines described in the ISO 9241, the different operation system based guidelines like Motif, IBM's *Common User Access* and the Apple *Human Interface Guidelines*, or the European Directive on office work with VDTs. These guidelines concentrate on general recommendations which are important for all users.

The expert view concentrates on general criteria such as consistency, feedback, language and terminology, error-handling, on-line help and dialogue styles. Best results can be obtained by using three or four evaluators. Each evaluator will only find a limited number of usability problems. But since different evaluators may find different problems, the total amount of problems found will be that much larger.

(3) User's view

Try to look at the application from the user's point of view. Take the viewpoint of different categories of users, as recognized in step 1. Examine aspects such as terminology, screen layout and dialogue structure. Here the developer tries to be an end user. There are no real users involved. The developer takes the user's viewpoint to critically examine his or her work. There are different methods of examination:

- evaluating the screen layout, which can be done by one person alone;
- using the application by trying to complete a set of predefined task, which is best done by two persons: one observer and one user;
- testing the interface with the help of specific, realistic scenarios.

A user's view is only possible if the developer knows who the users are. Taking a user's view has to be combined with a user orientation.

(4) User testing

Perform an interface prototyping session and usability test with 3–5 different users. Use evaluation techniques like thinking aloud, predefined tasks, interviews, exploration. User testing is one of the most important methods for usability evaluation. The system is being tested with 'real' users. There are different types of user testing such as iterative design and a formal usability test.

(5) Evaluation report

Write an evaluation report listing all problem items: the problem, an explanation, its cause, possible solutions, suggestions for further investigations. The report has two goals: the formal documentation of the usability evaluation and guidelines for future projects. The report should be easy to read, concise and concentrate on the evaluation results, both positive and negative. An evaluation report tends to concentrate on criticism and thus often has a negative tone, which in turn might offend the reader. For each problem the report should give:

- a description;
- a classification (annoying, severe, fatal);
- an explanation of both the problem and its cause;
- possible solutions;
- relation with other problems.

(6) Recommendations for improvement

To avoid repetition of errors the evaluation report also addresses some kind of learning goal: What did we do wrong, why did we do it wrong, how can we correct this error, and (one of the most difficult questions) how can we prevent these errors in the future?

Performing a usability evaluation is one thing; designing a good user-friendly interface is quite another. A usability evaluation focuses on problems, not on solutions. The problems which were found during the evaluation have to be translated into recommendations for improvement. This translation can be added to the evaluation project as an extra step. Extra, because finding solutions might require additional testing or discussion, especially since solutions can create new problems.

WHAT STEPS TO TAKE?

DIY consists of a minimum of two steps: performing an evaluation and writing a report. Step (2), the expert view, is the easiest step to take. The expert view is independent of the user characteristics and does not require any specific tool or situation. The expert view is applicable to paper descriptions, sketches of screen layouts, prototypes and ready-to-ship products.

Steps (3) and (4), the user's view and user testing, do require a previous orientation on users and thus have to be preceded by step (1). Evaluating an application by means of a user's view does not require any specific tool or situation. The user's view is applicable to paper descriptions, sketches, prototypes and ready-to-ship products.

User testing requires thorough preparation. The goals and method have to be defined; including a test environment, task descriptions for the users, and a method of registration. User testing is not possible with a written description of the system (the user has to be able to see the interface), but it is possible with sketches, prototypes and ready-to-ship products.

The development team should choose between steps (3) and (4); in most situations it is not very useful to perform both. If there is a coherent and easily accessible group of users and there is enough time and budget to schedule a user test, then this should be preferred. A user's view can be used if there is little time or there are no end users available for testing.

Step (6), recommendations for improvement, will often be part of the evaluation report. Most user tests will generate usability problems which need a specific survey on solutions. However, finding a solution often requires some extra testing or investigation. If there is a need for further investigations, separating the recommendations from the report will enhance speed and control.

This brings us to the last subject: finding the right moment for a usability evaluation. Evaluating a completed and ready-to-ship product would give you the best results, but the chances of your recommendations being implemented are fairly small. As the DIY method aims at system developers, the best moment for an evaluation is during system development. But how do the DIY steps fit into software development methodologies?

HOW TO FIT DIY INTO SOFTWARE DEVELOPMENT METHODOLOGIES

Here, we compare two different models for software development. The first model, 'traditional system development', discusses the incorporation of DIY in widespread

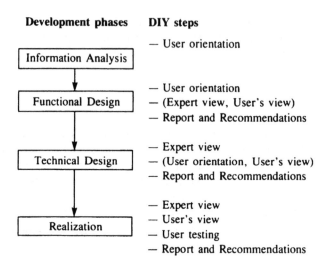

Figure 7.1   DIY steps in traditional system development.

system development methodologies based on a sequence of phases. As we will see, this model is not suitable for the effective use of DIY. The second model, 'iterative system development', concentrates on incorporating DIY into a new type of development methodology based on prototyping, joint application design and fast delivery, which are more natural partners of DIY.

Figure 7.1 shows the basic phases of a traditional system development project. The project is divided into four phases, which follow each other in sequence. Each phase requires its own expertise and thus different specialists. A phase starts with an orientation and concludes with a report. The actual product, the automated system, is developed in the last phase: realization and testing. During the phases information analysis, functional design and technical design the description of the system is being developed and specified in growing detail. Since the different steps follow in sequence, incorporating prototyping is not really possible.

A traditional system development methodology gives only limited possibilities to incorporate usability evaluation. Owing to the linear character of the project it is difficult to implement recommendations for improvement, especially if these improvements require redesign. The best opportunities for usability evaluation are during realization. Here the design 'comes alive'. Programmers can take an expert view or user's view while coding and testing their applications. Of course the user's view requires good knowledge of the users. This knowledge can be obtained by documentation or, even better, by visiting the users in their own environment. The realization phase also gives the best opportunities for user testing. There is a 'real' system which the users can work with, but it might also be too late. Since many projects exceed their budgets there is often not enough time and/or money to implement changes. A common solution is to ship the product and transform the problems to 'change requests' for the next release. In these situations the

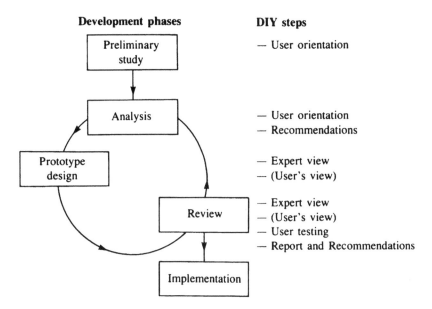

Figure 7.2 DIY steps in iterative system development.

recommendations for improvement are input to the analysis phase for the next release.

The best results can be obtained if the DIY steps are combined with a more prototyping and joint application design oriented development methodology like iterative system development. Here usability evaluation is a natural part of the development cycle and cooperation in multidisciplinary teams is far more accepted. Figure 7.2 shows the cycle of iterative system development. Following a preliminary study the total system is divided into a number of sub-projects, each of which should result in a complete, ready-to-ship subsystem. The different subsystems are developed one after another, thus giving the opportunity to implement changing requirements and to learn from previous experience. A small team of analysts, designers, developers and user representatives is responsible for developing a subsystem in a relatively short time. The whole team is involved with, and is responsible for requirements definition, prototyping, testing and feedback.

Usability evaluation is a natural part of iterative system development. The review and testing phase gives good opportunities to incorporate a user test and is a logical moment for evaluation. Despite the natural relation, usability evaluation is not a standard activity in all iterative system development projects. The prototyping, testing and feedback often focus on functionality and technical barriers, so that it is necessary to add explicit usability evaluation activities to the project.

The analysis phase should adopt a user orientation. In fact, the preliminary study already requires a user orientation, since users have to be selected to join the development team. But this first orientation is less detailed than the orientation during the analysis phase. During the prototyping phase all team members should use expert viewing while developing a prototype. The user's view may come at hand, though this is not really necessary since 'real' users are part of the team.

The review phase concentrates on a user test. The different parts and prototypes of the system have already been tested as part of the prototyping phase. The users involved in the project are not involved in the user test. There is always the danger that the user representatives get 'too involved' and thus lose their credibility. Successful review and testing leads to implementation. In this case the feedback concentrates on guidelines for future projects. If one or more serious problems are found during review and testing, serious attention has to be given to feedback. The recommendations for improvement can be part of the review phase or the next analysis phase.

CONCLUSIONS

In my experience, awareness of usability issues is growing. Within companies, prototyping and user involvement is becoming more widely accepted. Still, most developers would prefer to learn DIY and do usability engineering by themselves, rather than to hire a usability professional. Despite this urge to 'do it yourself' it is now widely accepted that usability professionals and interaction designers should be involved in projects requiring a graphical user interface or multimedia. It is difficult to prove that DIY has caused this change, but I'm sure it has helped to pave the road for more user-friendly interfaces and the acceptance of new development methodologies like iterative design.

REFERENCE

NIELSEN, J. (1990) Big paybacks form 'discount' usability engineering, *IEEE Software*, May: 107–108.

PART THREE

Field Studies

CHAPTER EIGHT

Service introduction and evaluation

JANS AASMAN, ENID M. MANTE and KARMA M.A. SIERTS

KPN Research, 2260 AK Leidschendam, The Netherlands

INTRODUCTION

For most telecom service providers the introduction of new technology was until recently almost entirely a technological affair. After the conception of the service idea (by technologists) and the preparation of the infrastructure to support the service (again by technologists) it was up to the poor marketers to find the customers to use the new service. Exaggerated? Maybe a little, but in general this is the main complaint of marketers.

So what goes wrong? In our opinion there are four aspects in the innovation and introduction process that are vulnerable to costly mistakes. The following paragraphs provide a brief analysis of these aspects.

Conception. The generation of new ideas is hard to manipulate. It is a matter of knowing the literature, listening to the market, and using the creative powers you have. There is nothing wrong with that. What is wrong is that we don't have a methodology screening for new service ideas. Such a methodology might consist of a model that systematically lists the variables and facilitators that play a role in the acceptance of new technology. We find that the general models in marketing science are inadequate to explain the adoption of new telecom services.

Selection of target groups. We are all aware of general misconception by technologists of consumer needs. In general they have an optimistic view of the innovativeness of the common citizen. Most people don't feel a need to incorporate new telecommunication devices and services in their daily practices and routines, and they often lack the technical knowledge and skills to operate them. Again it seems that we need a model that pictures the variables that influence the acceptance and usage of new telecom services.

Usability of new services. Gaining the acceptance of a new telecom feature is a one-shot opportunity. If it fails at the moment of introduction it may take years to repair the damage. By introducing a product or service that is too difficult to operate, that has inadequate functionality, or contains downright bugs you reduce considerably the chances of winning a larger share in the future market.

Evaluation. Although it is standard practice to evaluate the usage, and even the acceptance, of a newly introduced product or service, the results of these evaluations often prove to be of limited use, for a number of reasons. Both sales operators and marketing bureaux ignore the necessity to systematize the evaluation approach. Even if they are aware that longitudinal conclusions are necessary they think of new questions and new variables on the spur of the time-pressured moment.

For a product manager the only thing that counts is the selling of his or her product. However, there should be a company-wide awareness that every product is linked to other products in the past and in the future. The knowledge gained from different evaluations should be compiled in a form that can be used in future introductions and test markets. One way to ensure that this can happen is to use standardized instruments and measuring devices.

There is clearly a need to systematize the introduction and evaluation approach. The marketing and sales department of the Dutch PTT Telecom, in cooperation with the social science research department of PTT Research (ITB), has begun to develop such a systematic approach. In the following we describe a model of user acceptance, and then how this model is used to shape the introduction and testing of new services.

A THREE-FILTER MODEL OF USER ACCEPTANCE OF TELECOM SERVICES

To assess the future use and acceptance of telecommunication services, it is necessary to obtain information on the communication behaviour of individuals in society. What variables influence communication behaviour in general, and telecommunication behaviour in particular? How do domestic users react to new telecommunication services? What factors determine the acceptance process? To answer these questions a model is needed that gives insight into the enormous number of variables that govern communication behaviour, and the ways in which these variables are interrelated. Two things must be borne in mind:

1. Telecommunication behaviour is social behaviour, and is governed by socio-cultural factors. Although technology opens a wide area of possibilities for telecommunication, it is the human user who decides which of the possibilities will be adopted, and when.
2. The acceptance of new telecommunication services is a process very much like the acceptance of new technologies, of new working methods, and of new policy measures, and is subject to the same mechanisms.

The two points ask for a model that combines the variables which influence (tele)communication behaviour in general and the acceptance of new technology.

Social policy model

For the assessment of the factors which influence the acceptance of new telecommunication services we have translated the social policy model of Mayer and Greenwood (1980) into an individual evaluation process. The model consists of two parts: one part that can be influenced by the policy maker or, in our case, the telecom company; and another part that is beyond the direct reach of the policy maker or company. The part that can be influenced includes, for example, the choice of the service, the way it is developed and the way the implementation is planned and effected. The company is also able to choose the right facilitators; for example, the advertising campaign, the availability of a helpdesk, a user manual, the choice of certain role models, an attractive price or attractive added benefits. The limited space allowed here prevents us giving a detailed account of the model adapted from Mayer and Greenwood; see Mante (1994) for more details.

The rest of this chapter is dedicated to the second part of model, which describes the personal and contextual variables that influence the actual adoption process but which the telecom company cannot directly control. Unfortunately these are the very factors that have the greatest influence on the acceptance and adoption of new services and telecom devices. The crucial factor in this context consists of three basic variables: the need experienced by the individual, the technological infrastructure and the necessary technical knowledge that has to be available and finally, the presence or absence of an attitude that prevents the use of a specific service.

The three-filter model

To be able to put the right services on the market for the right people, marketers need to understand these three basic variables. Until now most attention has been given to segmentation along the lines of demographic and lifestyle characteristics; as long as these are strongly intercorrelated, this way of assessing consumer needs will suffice. In present society, however, individualization, democratization and internationalization have fragmented the classical divisions between social classes and age groups, making predictions about adoption behaviour highly fallible. Therefore it is necessary to understand the variables that govern (or used to govern) the intercorrelations between lifestyle and demographic variables; these are the contextual or situational variables in our model (see Figure 8.1). These variables can be described as three filters.

The first filter is formed by the personal characteristics and idiosyncrasies of the individual – his or her demographic position, knowledge and experience, attitudes and personal lifestyle – which influence personal preferences, and colour the choices he or she will make.

The second filter is formed by the situational variables: the structural way of life of the individual (for example, the household, the structuring of activities, social contacts, mobility); the specific communication situation (for example, is it purely

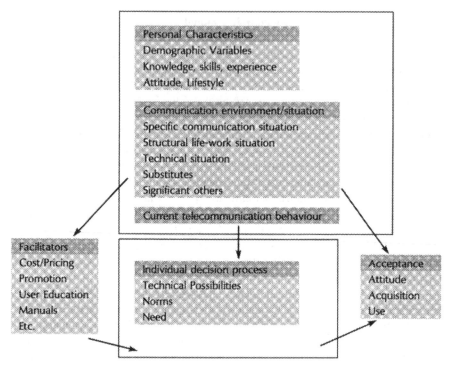

Figure 8.1 The three-filter model in relation to the decision process.

private, or more or less in a work context); the technical situation (the available infrastructure, other apparatus); the substitutes the individual has at his or her command; and the influence of immediate family members on the behaviour, norms and values of the individual. The situational variables determine the (manifest or latent) need for communication solutions and the way the individual is able to or allowed to make use of certain solutions.

The third filter is the present (tele)communication behaviour as an outcome of the individual and situational preferences. The present (tele)communication behaviour predicts to a high degree the (tele)communication behaviour of the individual in the near future. Note that the order of these filters is arbitrary, and will differ according to the kind of person or the type of service offered.

The individual decision process

The three-filter model is a model from a sociological perspective. It describes the (statistical) relations between the variables that determine the overall tele-communication behaviour and the acceptance of new technologies. The filter model also describes how these variables might influence the individual decision process involved in the acquisition and usage of new services.

Extensive desk research into the individual factors that influence the adoption of a new telecom service has led us to a simple, almost commonsense description of the individual decision process. This decision process itself consists again of three evaluation moments. In Dutch these moments are simply described as *willen, kunnen* and *durven* (to want, to be able, and to dare). *Willen* means: do people really want/need to obtain a solution for their communication situation?; *kunnen* means do they have the mental, financial or the technological ability to use the technological solution?; and *durven* can be best described as: do people have any moral or social inhibitions that prevent them acquiring or using the technology?

The resulting checklist

From this model we have derived a tentative checklist of the variables that influence telecommunication behaviour. This checklist is currently being subjected to testing. Hypotheses about the way the variables are interrelated are part of this testing phase. Testing has taken place (and is taking place) in expert interviews and in several test markets. Several multivariate analysis techniques enable us to single out the combinations of situational and personal variables that are most likely to affect the need for and the acceptance of certain services. We also added to the checklist a set of facilitators that influence the evaluation process. (Thus, for example, price is a facilitator in the evaluation of technical possibilities. The marketing mix influences the *dare* or the *need* dimensions; a handbook or a sticker might be a facilitator for the *can* dimension, etc.)

The variables of the checklist are operationalized systematically and result in basic questionnaires which are administered to respondents in test markets. The analysis of the data of our first test markets revealed that our questionnaire consists of several reliable scales (clusters of related variables). In addition, multiple analysis techniques revealed some very interesting clusters of telecom users that could be segmented according to the three-filter model characteristics.[1] Fortunately, a number of these characteristics give us an easy way to identify potential target groups in the market.

APPLICATION OF THE CHECKLIST IN FIELD TRIALS

We are currently in the process of validating and parameterizing our three-filter model and checklist. Validation is possible because we apply the checklist systematically in all our field trials. In these field trials eight steps can be distinguished:

(1) *Feasibility.* Besides the use of the checklist in generating questionnaires, the other important use of our checklist is to screen new service ideas in an early stage of the innovation cycle. For this purpose we have developed a tool that visualizes all the important elements and variables that play a role in the choice of a new service. With this device, by thinking through a new service, we are

able to eliminate several traps and pitfalls that can very easily crop up.

(2) *Choice of respondents/experimental subjects.* We use the segmentations resulting from our feasibility study to select the most likely subjects for pilots and field trials. In this stage of course we also select less likely subjects to validate the checklist. By using the model as a background for the selection we are also able to test hypotheses about acceptance and adoption by different groups of users. This enables us to be more confident the next time we put a service to trial.

(3) *Prototyping and usability.* In our usability tests the segmentations described in step (2) may be used. This enables us to adapt the application to the targeted user groups.

(4) *Choice of promotion campaign (facilitators).* The inclusion of the facilitators in our checklist ensures that in a promotion campaign no potential facilitator will be overlooked. Moreover, we are able to select likely facilitators based on the information obtained from previous test markets.

(5) *Experimental design.* The choices made in steps (2) and (4) determine to a large extent our experimental design. However, we may also use several types of promotion to measure the effects of facilitators.

(6) *Multiple evaluations.* As in any field trial we monitor the acceptance and the usage of the new service with zero, in between, and after measurements. Both quantitative and qualitative measurements are employed. The basic questionnaire is adapted to the specific service being tested, and is sent to all participants in the test market. During the test period, in-depth interviews may be held to assess attitude changes, complaints and other subjects of interest to the service provider. The influence of family members, personal lifestyles and attitudes may also be tested.

(7) *Analyses of automatic measurements from switching centres.* In addition to the evaluations we automatically log the usage of the service at the switching centre (of course with the users' consent). This gives us the 'hard' measurements we need to test the success of the service. We can also compare results of the 'hard' measurements with 'soft' subjective evaluations.

(8) *Comparison with other data (other test markets and periodic consumer research data).* The main goal of using a test market is of course to measure the potential of the new service and to identify problems that have to be avoided when introducing the service to the public. A second goal of using a test market is to identify the user groups or population segments that may be targeted at introduction time. The third goal is to validate and fine-tune our checklist. This will improve forecasts of the likely use and the likely user groups of future services.

CONCLUSIONS

We have found the approach described here to be a very fruitful one, for the following reasons:

1   The use of the checklist ensures a uniform approach to setting up field trials; it is the only way to accumulate information that may be used in the introduction of each new service.
2   The systematic combination of hard measurements of service usage with our consumer surveys and segmentation.
3   Close cooperation between the marketing and sales department and the social science research department of PTT Telecom ensures that the accumulated information remains available to the company.

NOTE

[1]   Our company does not currently allow us to provide a detailed account of the individual variables in question.

REFERENCES

MANTE, E.M. (1994) *The Telecommunication User Now and in 2015*, Paper for the COST 248 workshop, Lund, Sweden, April 1994.
MAYER, R. and GREENWOOD, E. (1980) *The Design of Social Policy Research*, Englewood Cliffs, NJ: Prentice Hall.

CHAPTER NINE

Evaluation of a multi-system situation

JAN LEGTERS

Department of Human Factors and Psychology, Hoogovens IJmuiden, The Netherlands

INTRODUCTION

Hoogovens IJmuiden is part of the Hoogovens Groep, situated in IJmuiden, The Netherlands. Hoogovens IJmuiden is an integrated steel works, in which all stages of steelmaking are performed, including tin coating, galvanizing and painting. Most of the process installation and equipment is run by operators from one or more of the 50 control rooms. While building the computer-controlled installations and control systems much attention was paid to the ergonomics of the user interface. In discussions with suppliers of the electrical and computer systems, including Siemens, ABB and General Electric, usability was a major item. The 1980s was a period of pioneering in this field, and Hoogovens IJmuiden played a substantial role in the development of user interfaces for complex control systems for the steel industry.

In a control room the main computer system is used for monitoring and controlling the installation and the process. Other computer systems are in use, for instance systems for planning and processing, for quality control, and for monitoring the condition of equipment. In the 1980s, when most of the systems were developed, Hoogovens itself could specify the user interface. Thus additional systems were adapted to the main system, and the usability of all systems together in one control room could be more or less achieved. Today, however, systems are usually bought as complete packages, i.e. actuators, hardware and software including the user interface, so that Hoogovens has little or no influence on the specification of the user interface. Such systems can be very usable, but in the control room, among other systems, this is not guaranteed.

In 1992 Hoogovens started a research project to study the usability of computer-controlled message systems, which is subsidized by the European Coal and Steel Community (Ergonomics Action Programme No. 6). Although these are commonly referred to as 'alarm systems', it is preferable to use the term 'message system', since the screens involved display not only warnings and alarms, but also messages and instructions for the operator. The project focused on message systems because they play an important role in emergency situations, when a lack of usability might easily lead to operator error. In an earlier ECSC programme the presentation of messages was the subject of research (Scholtens, 1993), whereas now the focus is on usability in conjunction with other systems used by the same operator. This means that the systems controlled are not all within one control room. Due to task rotation an operator usually works in more than one control room. Part of the project was to evaluate 16 systems, divided over eight control rooms.

THE EVALUATION METHOD

The method used is on-site semi-structured interviews and questionnaires. Operators are personally interviewed at their workstations. In 1993 16 systems were evaluated, spread over eight control rooms. The inquiry was aimed at detecting errors and irritations that arise due to confusing differences in user interfaces. It was hoped that the elimination of these differences would result in an improvement in quality.

The questionnaires used contained mainly multiple-choice questions with the possibility of adding remarks. The questionnaires were based on *Evaluating Usability of Human–Computer Interfaces* (Ravden and Johnson, 1989). In their overview of the method Ravden and Johnson describe how to use the checklist, although it was not possible to follow all of their advice. For example,

- A prerequisite for using the R&J checklist is that the evaluator actually uses the system to carry out the tasks it was designed to perform, as part of the evaluation. At Hoogovens the systems are housed in control rooms, not in laboratories, and most are quite complex. The operators (end users) had received intensive training in working with the systems, and so were interviewed at their workstations. They were able to give the evaluators an on-site demonstration of the system features.
- The tasks observed in the evaluation should be representative of the work to be carried out using the system, so that, in order to compare one system with another, the tasks should be representative to both. Those tasks are limited, because of the number of systems evaluated, and the differences between the applications. Standard tasks are data entry, finding information about an alarm, accepting and quitting a message. During the interviews more specific tasks were mentioned when necessary.

Table 9.1 lists the items in the questionnaire, with examples.

The members of the team of six experts on ergonomics conducted the interviews in pairs: one interviewed the operator, while the other observed the use of the system. Both could make remarks. The operator could read the questions from his own copy, and did not have to write. Depending on the answers or items, the operator was asked to demonstrate what he meant.

Eight control rooms were visited, in each of which two systems were evaluated by interviewing two operators from two shifts. In the second interview the

Table 9.1 Items in the questionnaire, with examples

Item	Example
Background information about system, user and control room	How many systems are used at the same time?
Development and introduction of the system	Does participation in the development of the user interfaces influence the operator's opinion?
Hardware	Are the screens monochrome or colour?
Operation of the system	What input device does the operator use: special keyboard, mouse, tracker ball, etc.
Data entry	Is it possible to stop data entry half way and continue later on without losing data?
Messages and alarms	Does the operator know how to react when alarms occur?
Consistency in the system itself	Is the same key used for the same function throughout the system?
Structure of screens	Is it always clear which screen is presented?
Linguistic usage	Does the linguistic usage link up with the oral usage?
Screen layout	Does the layout help in finding information?
Availability of information	Are other sources used to gather information?
Readability	Does the use of colour help to make the displays clear?
Documentation	Does the operator use documentation to run the system?
Preferences among different systems	Which input device does the operator prefer?
Consistency between more than one system	Is the same procedure used for the same task throughout all systems?

interviewers were interchanged. As mentioned above, all interviews had to be conducted in the control rooms while in operation. Because the ongoing process required the full attention of the operators, the interviews were often interrupted. Although such interruptions are a disadvantage because of the extra time they take, the interview method is the best way of gaining an impression of the actual conditions in a control room.

Because six interviewing teams produce a great deal of data, much attention was paid to proper presentation. All the information was entered into a computer database by one of the interviewers. He was assisted by his partner, while entering data, thus avoiding misinterpretations and inconsistencies in terminology. The

	control room 1				control room 2				
	System 1		System 2		System 3		System 4		
	S1	S2	S1	S2	S3	S4	S3	S4	
question 1									
question 2									
question 3									
question ..									

Figure 9.1 Layout of the evaluation document. Vertical: questions; horizontal: answers. S = subject (i.e. system operator).

layout of the evaluation document was designed to enable a comparison of the answers to each item; see Figure 9.1.

RESULTS

An analysis of the data showed some remarkable differences in opinions and in user interfaces, which have become the subject of further research. All items found were presented in a document.

A major issue related to the use of messages is that of the inflation of alarms. Additional sounds, suppression techniques or priority in messages are sometimes used. The presence of several message systems does not help the operator to solve the problem. A few systems were abandoned when an alarm appeared because of the mass of subsequent connecting messages. It is obvious that at such moments an unrelated alarm will go unnoticed. In alarm situations when response times have to be short, some systems slowed down owing to the large number of messages.

Differences between systems were found in the use of keyboards, for instance the use of F-keys. Also the use of colour was not consistent over all systems in one control room. Differences that might lead to errors with serious consequences were not encountered. Using two completely different systems was no problem, but the fact that almost identical systems differed in just a few crucial respects, sometimes leads to irritation of the operator.

For example, in one case three almost identical systems are positioned side by side in front of the operator, each with a VDU and a standard keyboard. All three systems are used for planning, monitoring and data entry for different tasks. The screen layouts are nearly the same. In two systems data processing starts after pressing the 'Enter' key, while in the third the data have to be confirmed by pressing 'Crtl-F1' before 'Enter'. Using just 'Enter' will have no response and will lead to delay of the total process and might lead to the loss of data. A second example is the way in which time is presented. The operator uses a database system and a process control system at the same time. The two systems are not connected to each other. Before the start of a process the operator chooses a process with an average time in hours and minutes. To optimize the time the operator checks the database system, which presents the correct time in hours and tenths of hours. The operator then has to convert the time to hours and minutes and, if necessary, adjust the process.

Besides the results of the evaluation, several remarks can be made about the method of evaluation. The method is very time-consuming; data acquisition for the 16 systems took about 96 hours, data processing 32 hours, and data evaluation

Figure 9.2 Control room at the hot-strip mill of Hoogovens IJmuiden.

about 108 hours. It is necessary to interview more than one end-user, and opinions are often quite diverse. With more than one interviewer there should be no doubt about the interpretation of questions. Otherwise this will lead to misinterpretations and worthless answers.

The idea is to have a method that is also usable for additional interviews at a later stage. This was done for one control room, four months later. It is no problem to do the interview once more, and the new data can be easily entered. The problem in this case was to draw conclusions while comparing the new items with the existing ones. One reason is the long period between the interviews: the details of the existing data were no longer readily available. The other reason is the restricted way answers and remarks were noted down; this provided insufficient information to identify precise details. When the interval between interviews is more than a few months the way answers and remarks are noted down needs to be more explicit and more explanatory.

CONCLUSIONS AND DISCUSSION

The evaluation

The enquiry approach used in this project gives a general idea of the problems involved in using more than one computer system in a control room. Besides that the handling of messages and alarms gives enough reasons for improvement, because of the large number of useless messages. The real influence of an extra system in one place could not be determined, so that a specific study had to be set up, in which the performance of one system was measured, while a second system interrupted the operator task.

The method

The method can be used in a multi-system situation to find general aspects of usability, rather than to find specific elements that need to be improved. To evaluate a new additional product, however, the results of this evaluation can help in the choice of relevant items, such as compatibility in handling messages and alarms, the use of keyboards, and the way of ordering and suppressing messages and alarms. Specific tasks can then be described. A short list of relevant items also reduces the amount of data and makes the evaluation more efficient.

The way in which the data are presented provides a good example for future evaluations, especially the flexibility to make new arrangements or cross-references. Therefore the answers and remarks should contain enough information to stand on their own merits. Answers like 'yes' and 'always' have to be avoided.

Testing two systems in a laboratory is probably a better way of determining the specific facts of what makes a user interface tolerate a second interface. In future it would be helpful if user interfaces could be categorized into compatible and noncompatible types, like hardware is now, or the use of user interfaces that are easily adaptable to each other.

REFERENCES

RAVDEN, S.J. and JOHNSON, G.I. (1989) *Evaluating Usability of Human–Computer Interfaces*, Chichester: Ellis Horwood.

SCHOLTENS, S. (1993) Presentation and treatment of communication and alarms on the screens of computer-controlled message systems, in: O. Berchem-Simon (Ed.), *Ergonomics Action in the Steel Industry*, pp. 323–32, Luxembourg: Commission of the European Communities.

CHAPTER TEN

Observation as a technique for usability evaluation

CHRIS BABER

School of Manufacturing and Mechanical Engineering, University of Birmingham, Birmingham, UK

NEVILLE STANTON

Department of Psychology, University of Southampton, Southampton, UK

INTRODUCTION

One could say that observation simply involves looking at something and recording what you see. The effort put into the development of usability laboratories would suggest that this definition has been taken literally; in some places, video-recording is performed in place of observation. By way of analogy, some students may spend an enormous amount of time and money photocopying papers, and appear to feel that ownership of the photocopy of an article represents, somehow, a knowledge of that article, and hence, removes the need to read it; so video-recording a person's behaviour could be seen as a means of recording it, without actually observing it. No doubt there are many usability laboratories which house hours of videotape to be analysed 'at a later date'. One of the reasons why this analysis often fails to materialize is the time cost involved in videotape analysis; we would estimate that 1 hour of recording could require around 10 hours of analysis. There are, however, approaches to this problem: either conduct the analysis 'live', or analyze for specific types of activity, or employ some form of computer support. Each of these approaches is considered in this chapter.

It was W.C. Fields who first stated what could be seen as a credo for human factors: 'Work fascinates me; I could watch it for hours'. However, there are many forms which can be taken when watching people at work, and just as there are many ways of watching paint dry, some are more informative than others. The topic of this chapter is 'observation', the process by which one watches work. However, it is not possible to separate the observation from the recording, the process by which one makes a permanent record of one's observation for consultation by another party or at a later time. When we observe something, what do we see and how should we ensure that other people will see the same thing? This is the principal question to ask. If our observations are being conducted as part of the design process, then it becomes important to collect observations which can be interpreted and applied by other members of the design team.

Even if our observations are conducted to get a 'feel' for what is going on, we ought to consider how to communicate this 'feeling'. We might decide to employ generalizations about a product, such as 'most people find the product easy to use', or we might prefer to provide some form of anecdotal evidence, such as one user of a ticket vending machine was observed rolling a five pound note into a small tube and attempted to insert this tube into the coin slot. While these statements and anecdotes may prove useful for informal discussion of product evaluation, one cannot imagine that they will play a major role in design practice. However, designers' knowledge of the people for whom they are designing is notoriously imprecise and fragmented, based on commonsense, anecdote and generalization. This is not to castigate designers; they often either have difficulty in finding sound, reliable, appropriate human factors knowledge for the product on which they are working, or have problems in applying such knowledge, either because of time constraints or because of other, competing concerns, such as manufacturability, appearance, materials, cost, etc. The point to note is that observation already plays a central role in design practice: observation of people using other products, observation of the design of other products, observation of the settings in which other products are used.

THE PROBLEMS OF OBSERVATION

We should aim to present a record which can be interpreted as unambiguously as possible, based on a set of observations which can be unequivocally defined. What are the consequences of not seeking unequivocal definition? The first is the problem of causality. If a person attempts to buy a ticket from a ticket vending machine and walks away from the machine without buying a ticket, to what extent is this due to the design of the machine, or to the knowledge of the person, or the situation in which the person is acting? We say that the person, in not buying a ticket from the machine, has 'failed', but let us suppose that the ticket office positioned near the machine has just opened and that the person leaves the machine to go to the office. Again, we ask does this reflect a problem with the machine or the person or the situation? A common problem which we find repeated in observational studies is the equation of an event with a cause, i.e. person X performed action Y because ... However, all that we can legitimately say from our observation is that person X performed action Y (and even then we may not be correct). We cannot legitimately say why the action was performed for the simple reason that we do not have access to the person's planning and justification for an action on the basis of simply observing them. To cope with the former problem, we may wish to record some commentary, made by the person on their action. This commentary could consist of a description of what they are doing, an account of their planning and intentions, and justification of what they have done, etc., all of which fall under the general heading of verbal protocol (see Bainbridge, 1990; Praetorius and Duncan, 1988), or could consist of discussions with another co-user of the product (Suchman, 1987; see also de Vries *et al.*, Chapter 17). While these approaches can provide additional data, they are felt to lie outside the remit of this chapter. A point worth noting, however, is that the use of verbal protocol may simply be a means of substituting one set of inferences concerning the cause of behaviour, i.e. inferences presented by the observer, with another set, i.e. inferences presented by the actor. This could imply that whether or not we supplement our observations with information gathered via verbal protocol, we are still faced with problems of inference, and hence, subjectivity. Thus, we will continue to claim that observation cannot objectively reveal people's intentions or reasons for acting in the way that they do.

A further problem arises from knowing what to observe and how often to observe it. While we might wish to limit our observations to the use of a particular piece of equipment, we are still left with the issue of who will be using it and where they will be using it. If we conduct our observations in a usability laboratory, can we be sure that it is sufficiently similar to the domain in which the object will be used? For instance, in one set of studies reported to us, the researchers found that performance times in the usability laboratory were consistently faster than in an office, and that perceived levels of workload for the same tasks were rated as higher in the usability laboratory than in the office. These differences were thought to stem partly from participants' perceptions of the usability assessment, and partly from the excising of the task from its normal work environment. This latter phenomenon led to problems of pacing the work, of planning tasks within the context of work routine and of the feeling of being watched.

Clearly, watching somebody can have a bearing on their performance. We may try to be as unobtrusive as possible, but the presence of an observer can lead people to engage in behaviour additional to their normal activity. For instance, observation could lead people to demonstrate a knowledge of how a product ought to be used, rather than how they actually use it. This is especially true if people are removed from the normal settings in which they use the product. Alternatively, one could attempt to observe behaviour covertly, i.e. by using hidden cameras. However, this raises issues of the ethics of conducting observations. Obviously, if the project involves a field study of several hundred people, it would be difficult to canvass all of the people to obtain their consent. For instance, a study of public behaviour on escalators in a shopping precinct involved recording the behaviour of some 1500 people over the course of two weeks. One could argue that if the observation does not seek to identify specific individuals, if the observation is covert, if the data from the observation are not used for purposes through which individuals can be identified, then one could relax the need to obtain permission.

A final type of observation which may be useful is known as participant observation. In this approach, the observer actually performs the task or work under consideration. For some studies, this may require a considerable period of training, e.g. when conducting participant observation of the behaviour of pilots in aircraft, and it may be easier to obtain similar subjective information from interviewing experts. However, actually performing a task can provide some useful insight into potential problems which users may encounter. If one is studying the design of automatic teller machines, a simple form of participant observation would be to use a series of different machines, noting the aspects of machine use which were problematic. This could then form the basis of an observation schedule for a study. Alternatively, if one is interested in the process by which products are designed, actually sitting in and contributing to a design meeting is an invaluable form of participant observation.

HOW VALID ARE THE DATA FROM OBSERVATIONS?

In simply observing people perform tasks in field settings, we have done little if anything to minimize the problems of causality. This stems from the difficulty that we have in exercising some control over the relationship between variables in such settings, and introduces the problem of error or bias. To a certain extent, we see what we look for when performing observational studies. Sometimes this bias could lead us to see events which have not occurred; usually it leads us to weight our interpretation of events towards a particular type. In order to produce valid data, one should be able to demonstrate that appropriate precautions had been taken to minimize the problem of bias. One could go as far as to say that whole panoply of methods and techniques developed by science exists as a means of minimizing the problems of 'human error' in observation. Naturally the rigours imposed by laboratory experiments not only restrict confounding variables, but can also restrict or otherwise influence people's behaviour. Experimental psychology has sought ways in which these problems can be overcome, and argued that properly collected data can be generalized to 'real-world' settings. However, one might exercise some caution as to the claim for generalizability. Thus, researchers are often caught between the Scylla of generating robust, generalizable data and the Charybdis of ecological validity, i.e. 'real-world' applicability.

We can answer the question by considering some of the different forms which the concept of validity can take. Construct validity can be defined as the degree to which variables accurately measure what they are intended to measure. It could be affected by confounding variables, which have not been included, or by recording error. If the observation is performed in an *ad hoc* fashion, with little or no attempt at producing a systematic recording scheme, then it is probable that construct validity will not be achieved. Thus, it is necessary to develop a set of tools and a protocol for using them. External validity refers to the extent to which one can generalize from the observations to all other classes. In laboratory experiments, this is ensured through careful control of variables. In observations, we need to ensure that a sufficient number of observations have been made on which to base our analysis. The definition of sufficient numbers is presented below. Internal validity refers to the relationships between variables, i.e. whether change in one variable causes change in another. But if confounding variables are introduced, then the observation may not be internally valid. This returns us to the problems of causality. Suppose we gave people an instruction booklet and asked them to read it until they had understood it. To what extent would measuring the time spent reading correlate with understanding? Even if we did find a positive correlation, we would not be able to say that more time spent reading led to more knowledge, or whether the people with more prior knowledge read more slowly to confirm their knowledge, or the people with less prior knowledge read quickly because they felt that they would not understand it. If we are confronted with such questions, it might be more sensible to consider some form of experiment.

Who to observe?

It is important to define the population which will be observed. Should one simply observe people that can use the product, or should one use people that are unfamiliar with the product, should one use a cross sample of the entire population, or people who are trained to use the product, or people who have been invited to use it, or people who are using it as part of their everyday life. Understanding that different people will use products in different ways can have a bearing on the selection of who we will observe and where the observation will take place. For instance, if one is interested in the ease with which people can guess the meaning of a specific label or the function of a control, then it will be important to use people with no previous experience of the product. If one is conducting field observations then it can be difficult to determine who will use the product; one is, to a certain extent, a victim of chance. However, through definition of a population sampling regime, it is possible to reduce the effects of chance. For example, simply setting quotas of people in different age groups may allow one to structure an observation sufficiently to maximize sampling.

Obviously, knowing who to observe does not necessarily mean that one will be able to say how many observations should be made. To some extent this might be constrained by access, time or cost pressures. If we assume that one is relatively unconstrained, it is possible to calculate the number of observations needed to produce data in which one can have a reasonable level of confidence. In the observation of users of ticket vending machines, the sampling was limited because of problems of access to the sites. However, given three days' access, it was decided to sample until no new errors appeared. This is crude. Is there a way to be more precise in specifying a sample size? If we assume that data collected in an exercise follow normal distribution, i.e. if plotting the data onto a graph can be assumed to produce an inverted-U curve (see Figure 10.1), then we can make assumptions concerning the distribution of the data. We assume that, given sufficient samples, we can say the data are normally distributed, i.e. that most instances of the event of interest will occur within the average value and that fewer events will occur as we move away from this value. If this assumption holds, then it

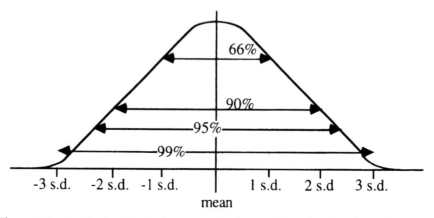

Figure 10.1 Hypothetical distribution curve, showing confidence levels and standard deviations around the mean.

is possible to define the number of samples needed to produce such a graph within certain limits of confidence, i.e. given a sample size, n, we can say that we can accurately describe a specific percentage of the graph. One might feel that this percentage should ideally be 100, but it is pragmatic to reduce this to levels of between 90 and 99%, as we can see below. If we maintain our assumption that the graph is an inverted-U, then we can determine equal amounts of deviation either side of the average value, i.e. standard deviation. We will not demonstrate the proof of this value in this chapter; suffice it to say that we can use this standard deviation to define the number of observations required, given the approximate distribution of some binary performance criterion, i.e. a criterion in which performance is defined as either X or not X.

We now return to the ticket machine study. We assume, from previous work, that the likelihood of success with the machine will be around 70%. This gives a likelihood of failure of 30%. Given some measure of standard error of proportion, and some assumed confidence level, we can calculate the number of observations needed. Let us assume that we require 95% confidence in the data, and that we are prepared to accept an error rate of 5%. Thus, error rate, $E = 5$ and from proof of the standard error of proportion, e_p (not given here), we can say that a confidence level of 95% can be determined by 1.96 of the error rate, e_p would equal $5/1.96 = 2.5$. We then use the following formula to determine the number of observations we require:

$$\text{No. of observations} = P_{\text{success}} \times P_{\text{failure}}/(e_p)^2$$

$$= 70 \times 30/2.5^2$$

$$= 336$$

Given that we recorded 340 observations, and that the actual success rate was 67%, this would suggest that we can assume that our observations will account for 95%

of the graph. If we were prepared to tolerate a higher sampling error, say 10%, then the number of observations would be reduced to around 80. Suppose that we wished to produce data with a 99% confidence level. In this instance, e_p would equal $5/3.3 = 1.5$. The number of observations would need to be increased to 933. Finally, assume that the probability of success was far higher than our estimate, say 95%. From the above equation we estimate that a sample size of 76 would be needed for 95% confidence, and 211 for 99% confidence.

What to observe?

Some forms of observation can be conducted in the absence of people actually using products. For instance, one could infer the volume of traffic in a public place from carpet wear; one could infer the use of functions on a machine from wear on buttons, e.g. a drinks dispenser had hand-written labels stuck on the coffee and hot chocolate buttons but a near pristine label for tea because the tea was awful; one could infer patterns of use from signs of damage or wear and tear, e.g. on a food mixer, a set of scratches and marks where the bowl was placed attested to the difficulty of actually putting the bowl onto its holder; finally, one could infer problems from the presence of additional labels on a machine, e.g. ticket vending machines with 'do not insert money first' (which shows a problem but could be misread).

These 'observations' rely on inferences about the states in which products operate, i.e. things which are done by the product. Such states can be recorded in terms of their frequency and duration, and additional temporal data, such as fraction of total performance time, inter-keypress times, etc. However, these data may be difficult to interpret without an indication of the actions performed by a user. These data can be defined in similar terms, but focus on human behaviour rather than product states. In either case, the simplest form of recording is a frequency count of specific events, i.e. product state changes or human actions. Table 10.1 shows a simple frequency count based on a study involving a prototype 'electronic book'. The product was a non-functioning mock-up, requiring an experimenter to perform physical actions, such as to turn a page, in response to 'user' commands. The commands were issued via buttons. The aim of the study was to assess the ease with which a user could guess the functions of the buttons.

In order for these counts to serve as data, it is necessary to consider the relationship between actions and states, i.e. to say which buttons were pressed to

Table 10.1 Frequency count for electronic book study

Turn to contents page	I I I
Turn to index	I
Turn to next page	ᵀᕼᑌI ᵀᕼᑌI I
Turn to previous page	I I

Table 10.2 Sample of field observations

Change given			Exact money		
Subject	Time	Outcome	Subject	Time	Outcome
1	16	S	8	35	A
2	23	S	9	56	S
3	48	A	10	21	R

produce the states. This required another frequency count. One could, of course, combine the counts into a single table. What these data do not tell us is the speed with which the product could be used (in this study, it was felt that timing the use of a mock-up would not produce realistic data). However, the point to note is that the actions were defined as unambiguously as possible, prior to observation. Table 10.2 shows data collected during a field observation of a ticket vending machine. Here, the actions are both timed and categorized. The aim of categorizing actions was twofold: to simplify recording and to produce information relating to a specific research question, in this case, how successful were users of the machines (S = success, R = success with repetition, A = abort).

There will be instances in which a number of items in a product will be used in combination, and the analyst may be interested in describing the frequency with which the products are used, together with the order in which they are used. From this, one may decide to reposition items within a product, placing the most frequently associated controls together. This form of representation can be achieved through the use of link analysis (see Figure 10.2). In link analysis, one indicates the number of times an item is selected and indicates, via lines, the frequency with which other items are selected next. Figure 10.2 illustrates the link analysis for a proposed redesign of a electron microscopy workstation using link analysis.

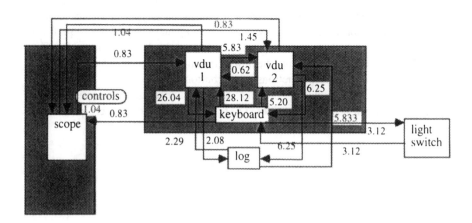

Figure 10.2 Link analysis of electron microscopy workstation.

Observation of 'live' performance is, in many ways, rich because it allows a potential to interact with the people performing tasks, etc. However, this level of interaction will depend on the degree of 'naturalness' one wishes. For instance, asking questions to participants may produce more information, but may alter or otherwise interfere with their task performance. Videotaping the task performance is useful in that it provides an opportunity for reviewing the performance, possibly allowing for more than one form of analysis, and can be edited for presentations. However, it is vital that the analysis be conducted with the aim of collecting some data. Software packages exist to allow analysts to sample a range of activities from recorded or live performance.

SOFTWARE PACKAGES FOR CONDUCTING OBSERVATIONS

Returning to the use of videotaped observations, we can see that videotapes can be analyzed using the simple techniques presented above. However, one of the reasons why people use video is to save time; the argument being that simply having a camera running does not require humans to spend time watching the behaviour, allowing the researchers to spend time on other projects and providing an opportunity to analyse the video at a later date. This last phrase is somewhat ominous; how late is the later date? If it is too long after the event, the recordings may be rendered useless. However, sitting through hours of videotape is time-consuming, and one feels that the process of sampling certain activities could almost be automated. While we do not have a fully automated 'intelligent' activity sampling computer, there are products on the market which allow quick and efficient analysis of video data (indeed, at least one of these products can also be used on hand-held computers for real-time observation). If we remember that the data collection concerns the frequency and timing of events, one could imagine software which uses keypresses to stand for specific events, with each keypress being time-stamped. This is the basis of packages called 'The Observer' and 'Drum'. Without going into too much detail, these packages have the potential to allow the analyst to control the rate at which a video plays, so the tape can be fast-forwarded to allow quick capture of rare events observed over long periods, and will perform basic statistical analysis of the frequency and duration of specified events. This allows analysis of videotapes to be performed almost in real-time.

CONCLUSIONS

Observation can provide illuminating insight into the difficulties people face when using products, both in terms of the types of problems and their frequency of occurrence. In order to collect usable data, it is important to determine the purpose of the observation exercise, and to ensure that there will not be an alternative approach. Observation can be time-consuming and, as Stanton and Baber (Chapter 24) suggest, can be substituted by error identification analysis. However, the time scale of observations can be reduced by employing software for the analysis. Observations do not require any specific human factors expertise, providing the observations follow certain basic guidelines, as outlined below.

Guidelines

The following guidelines summarize the points made in this chapter with a series of questions to be completed prior to an observation study:

1 Define clearly the aim of the observation study:
   'This study will examine the use of [which product?] by [which users?] in [which environment?] in order to collect data on [which aspects of product use?] which can be used for [which purpose?]'.
2 Define the scenario in which the product will be used:
   'The product will be used for performing the following tasks: ...'
3 Define the type of data to be collected:
   Will it be necessary to time activity or will a record of frequency of activities suffice? Could alternative methods yield the data in less time or with less effort?
4 Define the means by which data will be presented:
   Will it be sufficient to edit extracts of videotape to illustrate specific behaviour; is some form of descriptive statistics needed (how often, for how long, etc.); will further analysis be necessary?
5 Determine the amount of time available for observation.
6 Define recording tools; agree on the use of tools and ensure training/practice for all observers.
7 If possible, conduct observation in stages and meet to consider samples of data between stages; ask if what is being collected is useful, or whether a less time-consuming method could be used.

REFERENCES

BAINBRIDGE, L. (1990) Verbal protocol analysis, in: J.R. Wilson and E.N. Corlett (Eds), *Evaluation of Human Work*, London: Taylor & Francis.
PRAETORIUS, N. and DUNCAN, K.D. (1988) Verbal reports: A problem in research design, in: L.P. Goodstein, H.B. Anderson and S.E. Olsen (Eds), *Tasks, Errors and Mental Models*, London: Taylor & Francis.
SUCHMAN, L.A. (1987) *Plans and Situated Actions*, Cambridge: Cambridge University Press.

CHAPTER ELEVEN

Field-based prototyping

N.I. BEAGLEY

Centre for Human Sciences, DRA, Farnborough, UK

INTRODUCTION

Users require experience of a proposed product in order to be able to specify accurate and stable requirements (Luqi, 1989). Prototyping presents a concept in a tangible form, accessible to both the designer and the user. It serves as a tool to both stimulate ideas and refine them, through iteration, towards a product specification that fits the requirements of the user population. User requirements tend to change over the course of development, overtaking initial design specifications (Spence and Carey, 1991). The process of prototype evolution incorporating user involvement allows user requirements to develop in parallel with the prototype (Beagley *et al.*, 1993).

Prototyping provides an efficient medium for communication between the user and the designer (Benimoff and Whitten, 1989). At the early stages the prototype supports distillation and communication of the designer's understanding of the requirements. At the same time, the interface provides the user with an impression of the proposed application. A prototype can take many forms ranging from a sketch up to a fully functional simulation. The choice of prototyping approach depends on the product to be prototyped and the aim of the evaluation.

Software prototypes are most commonly developed as a specification tool and then discarded (Windsor, 1990). The alternative approach of evolutionary prototyping starts with an initial prototype that evolves through the prototyping process into the final product (Langen *et al.*, 1989). The use of prototyping tools to develop end products is increasingly evident (Poulter and Sargent, 1993; Shalit and Boonzaier, 1990). The large size and reduced speed associated with high level applications has become less critical due to steady improvements in processor speed and storage capacity. These limitations of prototyping tools are balanced by their flexibility and support for rapid development, allowing applications to evolve quickly into final products. The use of a prototyping tool allows the developer to concentrate on creating solutions, as opposed to overcoming programming problems (Jasany, 1990).

Prototypes are a useful method of communication within the design group. They display their real value, however, when applied to user testing. Testing, using a representative user population, is central to effective user centred design. Prototypes allow this testing at an early stage of development to avoid costly errors in specification.

By taking the users out of their normal working environment for prototype testing, there is a danger of missing significant usability issues. The user may be intimidated by a laboratory environment, modifying their approach to the system. In addition environmental conditions such as lighting or noise may impact on the overall operability of a system. It is important therefore that, where possible, the prototype should be tested in an environment that matches its anticipated operational environment.

Construction of a prototype requires the developer to make design decisions, decomposing the requirements specification to a set of design components (Tanik and Yeh, 1989). Prototyping supports innovation by providing a safety net, allowing the designer to test unusual design solutions. Components, proven through prototyping, can be developed further. Failed components can be replaced by alternative solutions in the iterated prototype. The developer must be prepared to discard components of a prototype as part of this process of iteration. The reduced development time associated with high level programming helps to enhance the disposability of prototype code, reducing the resistance to change.

A COMPUTER MAP PROTOTYPE

This chapter considers the iterative development of a computer map prototype. The prototype was designed as a disposable tool to assist in the requirements specification of computer maps for use in a military vehicle. A fully functional computer map prototype was developed to be tested in the field.

The task of navigating a military vehicle on the move cannot be isolated from the environmental influences. Effects such as vehicle motion, noise and operation in confined space and other parallel activities affect the task of navigation. A realistic environment cannot be easily simulated in the laboratory. The environment chosen for evaluating the computer map was the POD vehicle, based in the field (Figure 11.1).

The POD is a mobile, reconfigurable crewstation purpose built for human factors research. The POD's shell has been modified to include the vision systems present in a military vehicle. The task of navigation is made harder by the restricted view caused by operating with the hatches closed. The POD vehicle, used in the field, provided a highly suitable platform to test the prototype.

The prototyping tool chosen was SuperCard®, a high-level, graphically oriented programming environment. SuperCard® is a variant of the more common HyperCard® environment. The simplicity and learnability (Neilsen *et al.*, 1991) of these high-level languages made them suitable as prototyping environments by the group's human factors specialists. SuperCard® was chosen in preference to HyperCard® primarily because of its ability to support a large colour map.

Figure 11.1   The POD mobile reconfigurable crewstation.

Whilst important issues surround the physical design of the computer map unit, the scope of the simulation was limited to a consideration of the functional interface. Working within financial constraints, the prototype was developed and implemented on a desktop Macintosh computer. Although it may be desirable for a future map system to be repositionable or even hand-held, the available equipment limited the presentation of the prototype to a CRT screen.

FIRST PROTOTYPE

The starting point of any prototype is a concept. The decision to develop a computer map prototype was based on the assertion that a computer map would reduce a vehicle commander's workload. Whilst similar tools are used in some aircraft, the task differs considerably from military land vehicle navigation. Civil land systems available at the time of concept development were generally restricted to centralized tracking of fleet vehicles. The specific tasks of the user population made this tool individual. Consideration of these related systems informed the initial specification of the prototype without rigidly directing its design.

The population addressed by the prototype was the commander of a future reconnaissance vehicle. The group specified the core functions to be included in a computer map. These were considered to be: 1. Colour map; 2. Current vehicle position (overlaid); 3. Scrollable map; 4. Map marking. This specification was intentionally loose, providing a starting point to the prototype, without constraining creative design. Although it was possible for the group to specify their concept of a future map in greater detail this was avoided to allow the prototyping process to mould the direction of prototype evolution.

PROTOTYPE DEVELOPMENT

The group had previously used multimedia techniques for the evaluation of hardware panel design. Realistic buttons, switches, lights, etc. were incorporated on a textured background to produce a photographic representation of a panel, on screen. Animation and sound were used to provide feedback when the user interacted with the panel controls. This effect was further enhanced by the incorporation of a touchscreen for panel interaction. The advantage of this approach was that the panel could be quickly reconfigured, retaining full functionality. This three-dimensional approach was incorporated into the design of the computer map prototype. Control of the prototype was by 3D-effect software buttons, simulating hardware buttons. The group proposed that a real system should be controlled by hardware buttons. The flexibility of reconfigurable software buttons supported the flexibility of the prototype in support of iteration.

A 1:50 000 colour map was presented at the screen resolution of 72 dots per inch. The map controls were provided on software buttons surrounding a central map window (Figure 11.2). Size and labelling of the buttons was dictated in reference documents. The act of producing a functional prototype offered the means to tackle problems not covered in the literature, e.g. control of map scrolling in a vehicle environment. The prototype's solution was to place a large button at each side of the screen. Each button's position indicated the area to be revealed, e.g. pressing the top button moved the view to the area hidden above the displayed portion. With the absence of a prescribed solution to map scrolling, the act of proposing a solution in the form of a design component provided a starting point for evaluation.

Standard military symbols were used to mark the map. Selection of the map marking icons was controlled by the software buttons. A trackball was used to position the cursor over the map as opposed to using the touchscreen. Touching the screen in any area other than the software buttons had no effect. This was implemented to reinforce the impression of the software buttons representing their hardware equivalent.

FIRST TRIAL

The first prototype user trial was conducted as part of the 1992 POD field trial (Streets *et al.*, 1993). Representative reconnaissance crews were used for the trial. This served as a pilot trial to demonstrate the use of the equipment in the field. The desktop computer proved sufficiently rugged to withstand operation in a moving vehicle. The precautions taken included anti-vibration mountings and the

Figure 11.2   Initial computer map prototype.

prevention of hard-disk access on the move. The screen survived being used by the subjects as a rest for their coffee cups. The plan to provide vehicle location by beacon signal triangulation, for real-time positioning was abandoned due to insufficient accuracy. The computer map was simply used as a replacement to the paper map, providing no extra functionality. In this capacity it demonstrated the ability of the crew to interact with the system in a moving vehicle. In addition it highlighted interface issues to be addressed in the prototype's iteration.

PROTOTYPE ITERATION

Information gathered from the prototype's evaluation pointed towards areas that needed improvement. In addition, enhanced equipment became available between the date of the first field trial and the proposed date for the second.

The users occasionally displayed confusion when attempting to scroll the initial prototype's map due to the abrupt change of view. The iteration therefore took an alternative approach to the control of map scrolling. Smooth scrolling was implemented, tied to the metaphor of dragging a large map into the screen's 'window' (Beagley et al., 1994). The surrounding buttons used to control map scrolling were replaced by an active cursor. The cursor was displayed as an arrow that changed, depending on its position, to point towards the centre of the active map area. The arrow indicated the direction in which the large map would be 'dragged' when the trackball's button was pressed.

The ten buttons used to control the first prototype were reduced to four by limiting map marking to a generic target and a route marker. The functions of centring the map to the non-static vehicle icon and map decluttering were allocated to the two remaining buttons.

The availability of colour flat panels allowed the prototype to be presented on a less cumbersome, repositionable screen (Figure 11.3). The transfer to a flat panel produced knock-on effects such as the removal of touchscreen interaction. The trackball and button remained providing the only method of user control. The use of a flat panel, in turn, improved the range of acceptable viewing angle.

The first prototype demonstrated the principle of presenting a paper map on a computer screen. The function that separates it from a paper map, however, is the ability to overlay the vehicle's current position. Observation of crews using a paper map showed that the majority represent vehicle orientation by turning the map to show the vehicle heading at the top, whilst marking their perceived position with their finger. There were two possible solutions to providing this information for the map prototype. First, real systems such as a Global Positioning System (GPS), inertial navigation and an electronic compass could be integrated with the prototype. Alternatively, the information could be simulated using a Wizard of Oz technique (Wilson and Rosenberg, 1988). Feeding the vehicle's position and orientation in real time from the front of the vehicle provided a functional, reliable, low-cost system that was indistinguishable to the users from a real system.

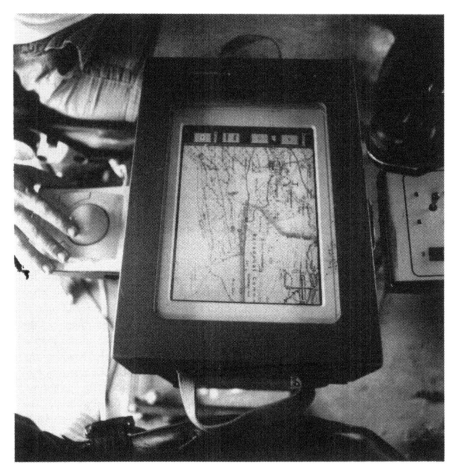

Figure 11.3   The iterated computer map in position within the POD vehicle.

SECOND TRIAL

The iterated prototype was trialled as part of the 1993 POD field trial. The revised prototype was fully functional and operationally reliable. The trial included a direct comparison between paper map navigation and navigation using the computer map prototype. The subjects consisted of eight pairs of representative reconnaissance vehicle crewmembers.

Measures used included observational data and applied questionnaires. The prototype incorporated background logging of all interaction with the prototype. The time based log could be used to subsequently 'replay' the session.

Operational measures including deviation from the route and decision time provided objective support for the benefit of a computer map as a navigational aid. The questionnaire and informal discussion, based on observation, proved to be rich sources of information. It is based on this information that the specification may be evolved further, through iteration of the prototype.

As anticipated, the computer map provided the vehicle commanders with an increased confidence in their position. This was due to the added information of position and orientation offered by the computer map. As a consequence more time was spent by the commander surveying the terrain. Due to the simplicity of the design, the crews experienced little difficulty in operating the computer map, despite minimal training. It is important to recognize the importance of simplicity of design in the stressful environment proposed. The usability of the overall system must be considered as the prototype evolves to include increased functionality.

Although not required for the trial scenarios, a requirement was specified by several subjects for a 'zoom-out' function. They considered a reduced scale map representation to be necessary for route planning as it would provide them with an indication of context. This is a good example of the subjects thinking beyond the tasks formally evaluated. They were able to consider effectively how they could use this tool to complete their regular job.

Presentation of the prototype on a flat screen allowed the user to reposition the screen. The user was able to reorient the computer map in the way observed when navigating with a paper map. With one exception, the subjects chose not to reorientate the computer map, despite reorienting their paper map in the comparison scenarios. An important area to be explored in the use of the map is to compare 'north up' map orientation with automatic 'vehicle heading up' orientation. A limitation of the prototyping technique at present is the restriction of processing power. Adding the function of map reorientation to the prototype, although possible, stretches the current power of the prototyping tool. The prototype developer must be resourceful in manipulating the prototyping tool to meet unusual specifications.

FUTURE PROTOTYPE EVOLUTION

The prototype is a valuable tool away from the field environment. It provides an object for presentation, stimulating debate amongst visiting groups. Senior visitors with a wide range of experience in operating past and present systems are able to bring a new perspective when considering the concept system. This debate generates ideas which in turn influence the direction of prototype evolution. An example of this is the consideration of the effect of the map as a communication node in the integration of the vehicle with its chain of command.

Areas proposed for future development of the prototype map include integrating a communication link and the use of new technologies such as hand-held devices, pen or voice input and remote sensors. The prototyping environments are evolving at a similar rate to the hardware improvements making it possible for the prototype to develop to meet the challenge of evaluating these advances in technology.

CONCLUSIONS

Applying a functional prototype in the field, using a representative population allowed the determination of recommendations for in-vehicle presentation of graphical map information. Over the course of development, the prototype evolved from a loose concept into a simulation demonstrating the predicted benefits of the concept tool. The considerations involved in introducing a concept tool into a complex system goes beyond the design of the tool. By using a functional prototype the user trial highlighted knock-on effects resulting from alteration of the overall system. Observation of these effects was made possible by the provision of a representative operating environment. The resulting scenario compares closely with the evaluation of in-service systems. The use of prototypes however offers a method of influencing the design process at a stage when alterations are less costly.

The strength of prototyping is its flexibility to evolve with the experience of both the designers and users. Creativity in design and disposability in construction supports the evolution of design to meet the functional requirements discovered through the iterative prototyping process.

ACKNOWLEDGEMENTS

This work was carried out at the Centre for Human Sciences, DRA, Farnborough (formerly A.P.R.E.). The views expressed are those of the author and do not necessarily represent those of Loughborough University, The Centre for Human Sciences, or Her Majesty's Government.

REFERENCES

BEAGLEY, N.I., HASLAM, R.A. and PARSONS, K.C. (1993) Hypermedia for the in-house development of information systems, in: G. Salvendy and M. Smith (Eds), *Human–Computer Interaction: Applications and Case Studies*, pp. 374–379, Amsterdam: Elsevier.

BEAGLEY, N.I., EDWARDS, R. and STREETS, D.F. (1994) Designing an electronic map simulation, in: S.A. Robertson (Ed.), *Contemporary Ergonomics*, pp. 144–149, London: Taylor & Francis.

BENIMOFF, N. and WHITTEN, W. (1989) Human factors approaches to prototyping and evaluating user interfaces, *AT&T Technical Journal*, September/October, pp. 44–55.

JASANY, C. (1990) Develop application programs in 120 days, *Automation*, February, pp. 34–35.

LANGEN, M., THULL, B., SCHECKE, Th., RAU, G. and KALFF, G. (1989) Prototyping methods and tools for the human–computer interface design of a knowledge-based system, in: G. Salvendy and M. Smith (Eds), *Designing and Using Human–Computer Interfaces and Knowledge-based Systems*, pp. 861–868, Amsterdam: Elsevier.

LUQI (1989) Software evolution through rapid prototyping, *Computer*, May, 13–25.

NIELSEN, J., FREHR, I. and NYMAND, H.O. (1991) The learnability of HyperCard as an object-oriented programming system, *Behaviour and Information Technology*, **10**(2): 111–120.

POULTER, A. and SARGENT, G. (1993) Hypermuse: A prototype hypermedia front-end for museum information systems, *Hypermedia*, **5**(3): 165–186.

SHALIT, A. and BOONZAIER, D. (1990) Hyperbliss: A Blissymbolics communication enhancement interface and teaching aid based on a cognitive–semantographics technique with adaptive predictive capability, in: D. Daiper *et al.* (Eds), *Human–Computer Interaction*, 499–503.

SPENCE, I. and CAREY, B. (1991) Customers do not want frozen specifications, *Software Engineering Journal*, July: 175–181.

STREETS, D.F., EDWARDS, R. and GOSLING, P. (1993) A mobile reconfigurable crewstation: POD report on proving trials, *APRE Report*.

TANIK, M. and YEH, T. (1989) Rapid prototyping in software development, *Computer*, May: 9–10.

WILSON, J. and ROSENBERG, D. (1988) Rapid prototyping for user interface design, *Handbook of Human–Computer Interaction*, Amsterdam: North-Holland, Ch. 39, pp. 859–875.

WINDSOR, P. (1990) An object-oriented framework for prototyping user interfaces, in: D. Daiper *et al.* (Eds), *Human–Computer Interaction*, 309–314.

Informal Methods

CHAPTER TWELVE

'Quick and dirty' usability tests

BRUCE THOMAS

Philips Corporate Design, 5600 MD Eindhoven, The Netherlands

INTRODUCTION

Testing provides feedback loops in the product creation process. This feedback can be provided throughout the process, through the testing of simulated concepts, working prototypes and products on the market. This chapter discusses usability testing within the context of the product creation process, looking in particular at how this is achieved in an industrial context. It draws strongly on work carried out by the Applied Ergonomics Group at Philips Corporate Design.

At the time of writing, 13 ergonomists are employed at Philips Corporate Design. We work as consultants for all Philips businesses, which means working in the development of products as varied as car radios, television sets, pagers, cellular telephones, electron microscopes, diagnostic ultrasound systems and shavers. The work of the group has been described by, for example, de Vries *et al.* (1993), de Vries and Thomas (1993), Thomas *et al.* (1991) and McClelland and Brigham (1990).

We offer services to clients throughout the product creation process from concept development to commercial release. These services cover both qualitative and quantitative user investigations, interactive development, generation of usability guidelines, design consultancy and evaluation, expert appraisals of product design, advice on standards and *ad hoc* ergonomics advice. The work of the group in usability testing is thus only one aspect of ergonomics effort applied to ensuring product usability.

The usability testing carried out by members of the group usually takes place in the usability laboratory located at Philips Corporate Design headquarters in Eindhoven. The functioning of this laboratory has been described by de Vries *et al.* (1994). The laboratory is shown in Figure 12.1.

Both formal usability tests and informal 'quick and dirty' tests are carried out at Philips. Quick and dirty tests provide a valuable opportunity to try out concepts at various stages during product development, as well as provide a means for usability testing where a formal test would be desirable, but is not feasible.

Figure 12.1   The usability laboratory at Philips Corporate Design.

FORMAL USABILITY TESTS

The quick and dirty tests can be described as 'stripped down' versions of formal tests. In order provide a basis for understanding where compromises are necessary, an outline is given in this section of the nature of the formal tests we carry out.

Formal usability testing at Philips Corporate Design is carried out to meet three main objectives:

- to discover particular user problems and potential flaws in design concepts,
- to determine how usable existing products are, with a view to setting baseline criteria for new development, and
- to determine whether predefined criteria have been met.

For all these objectives, measures of usability are required. Such measures are identified in ISO 9241, part 11.

ISO 9241 defines usability as: 'The extent to which a product can be used by specified users to achieve specified goals with effectiveness, efficiency and satisfaction in a specified context of use.' This definition of usability has proved to be a valuable aid to delivering a coherent message to our clients. It is also valuable as a means to specify the measures to be applied in usability testing. One of the main reasons for this is that the components of usability defined can all be expressed quantitatively, which facilitates comparison between products and product generations (see van Vianen *et al.*, Chapter 2).

Effectiveness is defined as: 'the accuracy and completeness with which users achieve specified goals'. It can be measured in terms of whether test participants are able to complete specified tasks. A distinction can be made between completion at the first attempt, completion on the second or subsequent attempts, completion with the use of a manual, completion with assistance from another person (e.g. the tester), and failure to complete. On the basis of these distinctions, a picture can readily be built up of how intuitive and how learnable the operation of a product might be.

Efficiency is defined as: 'The resources expended in relation to the accuracy and completeness with which users achieve goals'. Many investigators interpret this in terms of the time taken to operate a product (e.g. Macleod and Bevan, 1993), sometimes related to the number of errors made. Many of the usability tests we carry out are conducted using simulations or prototypes, and it is often uncertain whether the simulation response times are equivalent to those that would be achieved on the real product. An alternative measure of efficiency is to register all the individual actions a test participant makes in order to complete a task, including making errors and error recovery. In one test of a communications product, it was demonstrated by means of this measure that *adding* a key press to a particular critical task, which introduced more clarity in the interaction structure, dramatically reduced the total number of actions that the participants made in performing this task, that is they acted with more confidence and made fewer errors (see Thomas and de Vries, 1995).

We tend to measure satisfaction, 'the comfort and acceptability of use', on the basis of questionnaires using rating scales and qualitative interviews. These measures tend to be *ad hoc*, due to the lack of an off-the-shelf method which fully meets our needs. Some of the questionnaires we currently use are derived from Brooke's SUS (see Chapter 21). However, the validity of our derivatives has never been formally assessed.

Participants in our tests are selected according to profiles of product users. These profiles are usually supplied by the marketing personnel of the Philips Business Groups, and take into account such factors as income, product experience, age and gender, and level of technical interest.

A usability laboratory is an artificial environment, in that it can at best only imitate the environments in which a product might eventually be used, and the lack of ecological validity has been identified as a potential weakness of laboratory based studies (see, for example, Jordan and Thomas, 1994). However, no degree of realism will transform a room in a concrete block on an industrial site into, for example, a living room, and no simulation will fit into the environment the same way as a finished product. Nevertheless, when the salient aspects of the expected environment are reproduced in the laboratory, and when the tasks the participants perform are put into the context of a realistic scenario within this environment, a remarkable degree of realism and participant involvement has been achieved. In one test of a screen-based simulated telephone system, participants were asked to imagine they were working at the reception desk of a biscuit manufacturer and to deal with the company's incoming and outgoing telephone calls. Some of the participants identified so much with their assumed roles in this test that they even talked to the recorded voice in the simulation, fully expecting it to be a live person.

A particularly important aspect of carrying out formal usability tests is selecting the tasks participants are to perform. The tasks should be representative of all system levels, but should also be realistic in terms of the scenario presented during the test and of expectations of what parts of a product's (sometimes very extensive) functionality is likely to be used in 'real life'.

Generally, we are confident that our formal usability tests deliver what we require, although there is still a need for off-the-shelf measures of satisfaction and for improved data logging techniques.

THE NEED FOR A LESS RIGOROUS METHODOLOGY

Formal usability testing, as described above, is often time consuming and expensive. We work in an industrial context, and are not infrequently called on to carry out evaluations of concepts at short notice (one day is not unusual) to deliver a result in a week or less. In such cases, neither the time nor the budget will support full usability testing.

Less expensive and less time consuming methods for evaluation do exist, and are applied at Philips (see van Vianen *et al.*, Chapter 2). A commonly applied solution is to conduct an expert appraisal. Such an appraisal is usually carried out using a semi-formal structure based on a checklist such as that developed by Ravden and Johnson (1989; see Johnson, Chapter 20).

While quick and inexpensive, such evaluations lack some impact. It can sometimes be difficult for an ergonomist working in industry, simply on the basis of 'textbook' wisdom, to argue against the individual experience or opinions of other professionals, which have been built up over many years. On the other hand, it is difficult for other parties to contradict the experience of 15 or more participants in a formal usability test, especially when their difficulties can be presented in the form of a video summary during a project meeting.

Informal ('quick and dirty') usability tests are an effective compromise between the economy of expert consultation and the expense of formal usability testing. Such a test can be carried out with a few participants who, although they may not be drawn from the target population, are not technical experts and have no direct involvement with the development in progress. It can provide video clips illustrating problem areas, and showing real people experiencing those problems. It also draws on a variety of experience (as with formal tests) and can reveal potential issues and problems which an evaluation without user involvement will miss.

The use of an informal test need not necessarily be a compromise. If it is carried out in combination with a structured expert appraisal, an informal usability test can be a remarkably powerful but inexpensive tool for evaluating product concepts. This links the thoroughness of a checklist, the knowledge of the expert and the subjective experience of a few test participants. Evidence from these differing sources, when put together, can produce a convincing and visible picture of the strengths and weaknesses of a particular concept.

CONDUCTING 'QUICK AND DIRTY' TESTS

The 'quick and dirty' tests we carry out can be described as stripped down versions of the formal tests described above. They are particularly valuable during concept development, where they provide the possibility to try out a large number of ideas quickly and economically. Perhaps the greatest issue with respect to informal testing is how much of the formal test methodology can be sacrificed without losing validity or value.

An informal test can be structured to accommodate many of the aspects of usability discussed in the context of formal testing. Effectiveness can be measured on the basis of simple notes regarding whether tests participants successfully complete tasks. Satisfaction can be readily measured using standard questionnaires (limited though they may be). Thus, for effectiveness and satisfaction, results can be produced from the test in a comparatively short time, even in the absence of sophisticated logging and analysis tools. Similarly realistic tasks can be set to be performed in an appropriate scenario; standard tasks and even scenarios can be taken or derived from tests previously conducted (see also van Vianen *et al.*, Chapter 2). The major compromises made in informal testing therefore concern the measurement of efficiency and the recruitment of test participants.

Measures of efficiency, whether time taken, actions made or some other measure, often require extensive *post hoc* analysis. In theory, this analysis time can be shortened through the use of automated logging and analysis tools, but we have so far failed to find a tool effective enough to provide useful results within the tight time constraints which dictate the use of informal testing. Nevertheless, a qualitative description of the problems participants have in performing tasks may be considered to be sufficient for indicating whether or not the participants are having to do too much or are experiencing excessive frustration in manipulating a particular product.

Recruitment of participants, even for formal testing can sometimes pose major problems, particularly when a very specialized user profile is required. To find test participants with particular professional backgrounds, and sometimes even to match profiles for a more general (mass) market, requires time, and of course sufficient budget must be available for recruitment. If a test must be carried out within a period sometimes as short as a single day, recruitment of suitable participants becomes a major hurdle.

In order to carry out an informal test, it becomes necessary to abandon a strict match to the target profile and to make the best match possible with the people available. These people will in most cases be the staff at the site where the test is to be carried out. In the 1960s and 1970s it was said of the study of psychology that it was the study of male undergraduate psychology students; to be unkind, it might be said that informal usability testing is the study of the reaction of secretaries to product design. If informal tests are carried out frequently, some of these people will develop a degree of test sophistication even greater than the 'regulars' from the subject pool.

Non-representative participants with experience of product testing will have a different view of the product from that of a user using the product in the real world. However, the use of sophisticated participants is not an entirely unmitigated evil. We have found that these participants are much more comfortable in the test environment, they are more forthcoming in their comments (especially when they are personally well acquainted with the tester) and they are often more prepared to be critical of poor design.

An important issue when using non-representative test participants is whether the problems they experience or even the points that delight them will be the same as those of the target population. This issue has yet to be satisfactorily resolved. Representative participants are more likely to be aware of operating practices or nuances for particular products or applications. On the other hand, sophisticated participants are likely to be more sensitive to usability issues when given a brief exposure to a product or product concept. Ultimately, irrespective of background, any test provides the opportunity to draw on a range of experience, rather than relying on the judgement of a single expert.

An informal test, by its very nature, is limited in the data produced. Although general statements can be made regarding, for example, the number of test participants completing a particular task, it would be very questionable to submit such data to more extensive statistical analysis, if only because variations in the test population will almost certainly be different from those of the target population. The results of an informal test are illustrative rather than definitive, that is they can reveal potential strengths and weaknesses, but should not be used to measure differences.

One of the greatest difficulties in carrying out an informal test is simply finding something to test. If budget is not available to carry out formal usability testing, then it is unlikely to be available to build simulations or prototypes. Thus, informal testing would appear to be useful only for examining an existing product. Testing, particularly low cost testing, however, is probably most valuable during development, so that feedback from potential users can be taken into account in further design iterations. Some feedback can be obtained from simple paper based mock-ups, but our experience has shown that the comments and criticisms obtained by such means tend to be somewhat superficial – really useful test results do seem to depend on being able to represent the dynamics of an interaction.

In some cases, simulations are made for other purposes than for usability testing. This can be for purposes of demonstration or display. A simulation provides

- a point of focus for the development team,
- a means for the transfer of knowledge, and
- a means to ensure that iterations take place in an agreed direction.

Where a simulation is made, testing can also be carried out. Often product managers, ergonomists and others working in a development team must rely on complex technical specification to judge the adequacy of a concept. Sometimes even obvious problems can be missed if this is the only medium with which an interaction concept is represented.

A fundamental question to be raised with respect to informal testing is: when does 'quick and dirty' become 'fast and filthy', that is, when do the compromises made become so extreme that the method ceases to be useful? This is essentially a question of validation. In our experience, the results of informal testing complement the results of expert appraisals, which suggests that the approach does have some validity. They do highlight important strengths and weaknesses in early concepts and serve to provide a broader viewpoint of a product or concept which is not coloured by a more technical understanding. In some studies, as few as two people have participated in informal testing, but this is perhaps too limited; five or six participants provide a broader scope, both for getting different points of view and also for generalizing the results. Some investigators (e.g. Virzi, 1992) argue that as few as four or five participants provide adequate results even for formal usability testing. Although the debate concerning test participant numbers continues, at least it would seem that low numbers are acceptable for the less rigorous demands of informal testing.

CONCLUSIONS

Experience at Philips has shown that informal 'quick and dirty' testing can be an effective technique to evaluate the usability of a product, especially if this is done in combination with a structured expert evaluation. The main compromises which have to be made, compared with more formal usability testing, concern measures of efficiency and, more critically, the recruitment of suitable test participants. A further problem may simply be the lack of a dynamic simulation or prototype with which to conduct the test. Our experience suggests that a test with five or six non-representative test participants will reveal major strengths and weaknesses in the usability of a product. They complement the results of expert evaluations, which suggests that the approach has some validity. Finally, they can add valuable insights and protect against the potentially limited view of a single evaluator.

Far from being an undesirable technique to be used only as a last resort, informal testing provides an opportunity to try out new or different ideas economically and frequently. It allows the ergonomist to obtain corroborative data at little extra cost, and it provides a focus with which the dynamics of an interaction, as seen from a users' point of view, can be made visible. Informal testing is not necessarily a substitute for formal testing; it is a valuable supplement, particularly in early concept development when formal testing might not be appropriate.

REFERENCES

ISO 9241, *Ergonomic Requirements for Office Work with Visual Display Terminals (VDTs), Part 11: Guidance on Usability Specification and Measures*, ISO DIS, 9241-11: September 1994.

JORDAN, P.W. and THOMAS, D.B. (1994) Ecological validity in laboratory based usability evaluations, in: *Proc. 38th Annual Mtg of the Human Factors and Ergonomics Society*, pp. 1128–1131, Santa Monica: Human Factors and Ergonomics Society.

MACLEOD, M. and BEVAN, N. (1993) MUSiC video analysis and context tools for usability measurement, in: S. Ashlund, K. Mullet, A. Henderson, E. Hollnagel and T. White (Eds), *INTERCHI '93 Conference Proceedings*, p.55, New York: ACM Press.

McCLELLAND, I.L. and BRIGHAM, F.R. (1990) Marketing ergonomics: How should ergonomics be packaged? *Ergonomics*, **33**(4): 391–398.

RAVDEN, S.J. and JOHNSON, G.I. (1989) *Evaluating Usability of Human–Computer Interfaces: A Practical Method*, Chichester: Ellis Horwood.

THOMAS, D.B. and DE VRIES, G. (1995) *A User Centred Approach to the Design of Business Telephones*. Paper presented at the 15th Int. Symposium on Human Factors in Telecommunications, Melbourne, Australia, March 6–10, 1995.

THOMAS, D.B., McCLELLAND, I.L. and JONES, D.J. (1991) Ergonomics and product creation at Philips, in: Y. Quéinnec and F. Daniellou (Eds), *Designing for Everyone: Proc. 11th Congress of the International Ergonomics Association*, pp. 1070–1072, London: Taylor & Francis.

VIRZI, R.A. (1992) Refining the test phase of usability evaluation: How many subjects is enough? *Human Factors*, **34**: 457–468.

DE VRIES, G. and THOMAS, D.B. (1993) Ergonomics and consumer electronics products: A report from Europe, *Common Ground*, **3**(3): 1, 12–13.

DE VRIES, G., THOMAS, D.B. and McCLELLAND, I.L. (1993) De Applied Ergonomics Group van Philips Corporate Design (The Applied Ergonomics Group at Philips Corporate Design), *Tijdschrift voor Ergonomie*, **18**(5): 22–25 (in Dutch).

DE VRIES, G., VAN GELDEREN, T. and BRIGHAM, F.R. (1994) Usability laboratories at Philips: Supporting research, development and design for consumer and professional products, *Behaviour and Information Technology*, **13**(1): 119–127.

CHAPTER THIRTEEN

Effective informal methods

GARY DAVIS

Gary Davis Associates Ltd, Baldock, UK

INTRODUCTION

As an ergonomics consultancy group, Davis Associates undertakes projects that often involve the evaluation or development of a user interface, be it a computer-based software product, or a product-based control/display interface (for convenience, these will be collectively termed 'products'). However, we are quite often asked to contribute within short time-scales and tight budget constraints. We find ourselves having to accept these constraints because, if we did not, either the project would progress with no user-centred input at all, or, one of our competitors would be handed the work. This is the background to our application of informal methods for usability evaluation.

The methods we use are not new or sophisticated, yet they are effective. We will usually involve small numbers of existing users – where they exist. Our methods usually result in significant improvements to the usability of existing products, or the development of new products with minimal usability problems.

EVALUATING EXISTING PRODUCTS

In the evaluation of existing products, it is the acquisition of data that can cause most difficulty in terms of time-scale and cost. The following are the four data acquisition methods we most commonly adopt when evaluating an existing product:

- talking with existing users,
- user questionnaires,
- observation of existing users (usually in combination with talking),
- 'expert appraisal' of the product.

Talking with existing users

When attempting to evaluate the usability of an existing product, the most valuable sources of information are the existing users. In most cases, the existing users have experienced the best and worst aspects of the product and all we, as usability evaluators, have to do is extract their experiences and learn from them. Of course, this is never as easy as it sounds, because the users are (usually) human beings, and as we know, human beings are extremely adaptive and can learn to cope with the worst interfaces imaginable. In many cases, they will often have forgotten how incredibly difficult they found the product when they first encountered it. In other cases, users will be justly proud of their mastery over the product – an achievement which boosts their self-esteem and perceived status within their organization.

It is essential to read through the users' virtuosity and overcome their occasional reluctance to share their hard-earned experience. This can best be achieved by understanding the user's task related goals and by pointing out examples of where those goals are being compromised by the existing system. It is essential to establish a rapport with each user on a personal level; to acknowledge the skills they have developed; and to encourage them to critically assess the existing system.

In many cases, we, as evaluators, are initially unfamiliar with the product and discussions with the users are usually the most expedient way to become familiar with it. These discussions are usually video/audio-recorded enabling review at a later date. It is rarely possible to learn the entirety of complex products in this way, but this does not matter if the more significant usability problems can be identified with only a superficial knowledge. For example, when we were evaluating the usability of avionics panels for Airship Industries, we did not have to become airship pilots in order to specify the optimum configuration for the avionics control interface.

The number of users available for the evaluation process is often limited. We would normally hope to talk to about a dozen users; sometimes we are forced to make do with only one; more often, we have access to four or five. In fact, it is our experience that using more than four or five produces diminishing returns in terms of additional relevant information. This conclusion was also made by Virzi (1992) who found that four to five users usually highlight about 80% of a product's usability problems. Given the constraints within which we usually work, 80% is an acceptable level of success.

User questionnaires

In most cases, we have found that questionnaires are less desirable than a direct structured interview. Often, in our work, the time required to prepare a suitable questionnaire is disproportionately large in comparison to the number of subjects to be interviewed and/or the budget available. What is more important is that the flexibility of a one-to-one interview is far greater than any questionnaire, and allows a rapport to be established with the user. Having established a good user rapport, the evaluator is likely to uncover many more usability problems than would have been the case with a formal questionnaire. Therefore, we mainly use questionnaires only when:

- we need to acquire responses from a larger group of users (such as our survey of over 2000 company car drivers);
- it is impractical to be with the users for the appropriate period of time (for example, during a full working day);
- we and the users do not share a common language (for example, we translated a questionnaire into Cantonese when we required detailed feedback from Hong Kong train drivers about a cab mock-up).

Observation of existing users

The third main method that we apply is the passive observation of users. This will sometimes be done directly but, more often, is video/audio-recorded for subsequent analysis.

There are often practical problems to be overcome, such as how to best capture the user's behaviour and the product's responses without affecting the user's normal behaviour. The advent of miniature video cameras allows them to be located very discreetly, and very soon after being told of their presence, users often forget all about them and carry on as usual. We usually resort to multiple video cameras with subsequent detailed video analysis. Analysis of the video recordings is not usually rigorous in terms of quantifiable data. Usually, qualitative information is of far more interest, highlighting issues for further discussion with users.

'Expert appraisal' of the product

As virtually naive users of a product, we cannot hope to interact with it as readily or extensively as the experienced users. However, hands-on experience with the product enables the identification of further usability problems that may have been overlooked during user discussions or user observation.

This method is even less structured than those described above, and it has to be said, the results are less consistent. The method is more reliable if the evaluator has considerable experience of this 'seat-of-the-pants' approach.

EVALUATION DURING NEW PRODUCT DEVELOPMENT

Our work in the development of new user interfaces involves more rigorous evaluation methods towards the end of the development cycle, once prototypes and working versions of the products are available. During the earlier mock-up stages however, we find that informal methods are more appropriate, enabling usability problems to be resolved before committing to software or hardware that may be expensive or impossible to modify at a later date.

Two typical informal methods applied during the early stages of product development are: evaluation of paper-based mock-ups, and evaluation of software-based mock-ups.

Review of paper-based mock-ups

When developing new products, we do not have the benefit of existing users from whom we can learn, or an existing product that we can evaluate. It is therefore essential to produce something tangible that can be reviewed by ourselves, the client and by future users as early as possible in the development process. With items of 3D design, early production of block models greatly assists development. Similarly, by using very simple paper-based mock-ups of user interfaces, we are able to make early judgements about the most suitable style of user interface.

In one project for the Royal Mail (the design of a user interface for a system controlling an automated sorting machine), we were able to use these paper-based representations of windows and other screen elements to confirm our belief that a graphical user interface was indeed the most appropriate approach. In this case, the users were management grade, but most were not computer users and few had experience with computer packages of any kind. We provided the users with a paper-based metaphor of a software-based interface, which itself is a metaphor of the real sorting machine. Complex as this sounds, from the users' point of view it is very straightforward and appeared to cause them no difficulty in interacting with this 'virtual' interface.

Our users were able to manipulate the screen elements on the desk top and, through doing so, clarified for us and themselves the true goals of the task. For example, as they moved the various postcode blocks around on a table, it became apparent that resorting, inserting and deleting of postcodes were fundamental functions. The new interface had to allow these functions to be performed in a flexible and intuitive manner so that the users could concentrate on meeting their goals rather than on the use of the system itself.

Evaluation of software-based mock-ups

At a further stage of development, screen mock-ups are put together primarily in order to assess the graphics, legibility, colours, etc. We have successfully used these non-functioning screen mock-ups to gain feedback from future users by simply describing their functions.

The screens can also be animated by linking several fixed screens with functioning 'buttons' (using packages such as 'Hypercard' or 'Macromind Director'). This technique provides the users with a feeling that the system is fully operational and they are therefore able to provide more valuable feedback. These active mock-ups can be very elaborate and, if run on a suitably fast processor, can emulate prototype software.

CONCLUSIONS

In the experience of Davis Associates, informal methods in the evaluation of existing products and in the early development of new ones have proven to be as valid as more formal methods. In fact, informal methods can have distinct advantages in terms of greater speed and flexibility, and lower cost. However, informal methods are limited in the depth of analysis that can be achieved – the results are usually not statistically valid or repeatable and, because of their less structured nature, there is a greater reliance upon the experience and skill of the usability evaluator.

REFERENCE

VIRZI, R.A. (1992) Refining the test phase of usability evaluation: How many subjects is enough? *Human Factors*, **34**: 457–468.

CHAPTER FOURTEEN

Getting the most out of 'quick and dirty' user interface simulations

ARNOLD P.O.S. VERMEEREN

Delft University of Technology, 2628 BX Delft, The Netherlands

INTRODUCTION

Much of the recent literature on human–computer interaction emphasizes the usefulness of conducting user evaluations early in the user interface design process. Also, many of the user interface designers themselves see the need to get feedback from users during the design process (e.g. Bekker and Vermeeren, 1993). In this chapter we focus on user evaluations in a 'concepts and sketches' design phase. In such a phase, designers conceptualize basic design ideas and try to come up with alternatives. In later phases, designers elaborate upon these ideas and try to combine and detail them to create a complete and coherent design.

To get feedback from users on design ideas, designers need concrete material to confront users with: sketches, animations, storyboards, verbal descriptions, etc. The question is: what materials are appropriate in this phase? For evaluation of usability aspects it is not enough to passively present materials to users; at least some degree of interaction with (simulated) parts of a design is required. For building simulations that users can actively use, one needs to anticipate users' behaviour. Therefore, creating such simulations requires more effort than building pure presentation materials. Full anticipation of users' behaviour would require fully functional and realistic prototypes. As design projects are usually under heavy time constraints, it takes too much effort and time for designers to build such prototypes in a 'concept and sketches' phase. Designers have to make trade-offs between the effort and time they invest in making simulations, and the amount of interactivity, functionality and representation accuracy that is needed for valid and useful evaluation results. Unfortunately, little is known about how left out or inaccurately simulated design elements may distort evaluation outcomes. Therefore, designers have to make such trade-offs based solely on their intuition and experience. In our research, we search for information that can help designers in making informed choices on this matter. Such information can help designers to find out how to minimize their efforts of building simulations and still get user feedback in a sound and efficient way. In this chapter we describe a strategy that can help designers to determine what design features can best be evaluated and to derive requirements for simulating those features. First, we explain the stepwise approach to the strategy. Subsequently, we describe the 'roles' that form the core of the strategy, and finally, we illustrate the use of the strategy with a realistic example.

A STRATEGY FOR DERIVING SIMULATION REQUIREMENTS

Before describing the proposed strategy, we would like to stress again that in a 'concepts and sketches' phase, designers creatively try to come up with a range of design ideas and are not yet actively trying to synthesize these into one complete and coherent design. In this phase, designers usually do not conduct user evaluations for measuring the effectiveness, efficiency and user satisfaction of their design solutions. They merely want to get a feel of what design decisions to make, by observing users who struggle with their design ideas.

Obviously, requirements for a simulation are largely dependent on the design features that are evaluated, as well as on the usability aspects for which they are evaluated. For example, a paper-and-pencil simulation will not be very useful in evaluating the convenience of procedures for manipulating lines in a drawing program, but may be very useful in investigating guessability of icons in a public information system.

The strategy we propose can be described as a stepwise action plan:

1. identify the design features that most need evaluation;
2. analyze the identified features for the 'roles' they play in interactions;
3. select, for each identified feature, the 'roles' that most need evaluation; these roles will guide the evaluator in deriving requirements for the simulations.

1. Identifying design features that most need evaluation

In many cases, designs carry many conventional, well-established design features in them. Usually, only a limited number of design ideas are so new, or so critical to the design that the designer feels the need to evaluate them in the 'concepts and sketches' phase. The strategy suggests focusing simulation and evaluation efforts on those new or critical design features, and trying to simulate the complete design in this phase. For example, one could choose to evaluate only a specific metaphor or one specific function.

2. Analyzing the features for their roles

One single part of a design feature can serve to play a number of different 'roles' in an interaction. For example, one part of a function (e.g. the icon representing it) could be meant to inform users about the effect of using the function. In that case, the role is one of explaining to new users something about the function. Other parts of the function (e.g. the action procedures and mouse actions that are needed for applying the function) could play a different role. For example, making convenient and efficient application possible for experienced users. Step 2 of the strategy suggests analyzing the features selected in step 1, for the roles its constituents can play in the interaction.

3. Identifying the roles that most need evaluation

In step 3 the designer identifies the roles for which the selected features will be evaluated and for which simulations have to be built. Different roles will generally lead to different simulation requirements. For example, a designer could choose to concentrate simulation efforts on the role of informing new users instead of on the role of supporting convenient use by experienced users. In that case, those parts of the feature that are important for playing that role need realistic simulation (here, this could be the visual appearance of the icon and the effect one gets after using the function). Other parts of the feature (e.g. the procedures for applying the function) could do with a less realistic simulation in that case.

As the concept of 'roles' is at the core of the strategy, we will now describe four 'key' roles we have identified.

INTERFACE ELEMENTS THAT PLAY THE ROLE OF ...

The roles are inspired by the 'metaphors' and 'paradigms' for designing user interfaces that are described by Verplank (1989), as well as by the 'perspectives' described by Bødker (1991). We have identified the following roles: 'routing signs', 'tools', 'media' and 'intelligent assistants'.

Routing signs

On roads and in large buildings, routing signs serve to inform people about where they have to travel to arrive at a desired destination. Similarly, interface features may have as their main role to explicitly inform a user where to find certain functions and what procedures to follow in applying the function. For the proper functioning of elements as 'routing signs', aspects like consciousness and meaningfulness of command names, icons and status information are of major importance. Thus, simulations that are meant to evaluate features for this role should be accurate in simulating especially those aspects. In many cases, this can be achieved by using paper-and-pencil simulations of screens or panel layouts.

Tools

In everyday language, the word 'tools' is used to denote the artefacts that people use to create or modify something. In normal use of tools, people focus their attention on applying learned skills and strategies for modifying or creating objects or information and not on how to use the tool itself: when writing with a pencil, one concentrates on formulating sentences and preferably not on how to hold and manipulate the pencil. Many interface design features are designed for a role similar to that of conventional tools. In this context, the materials and objects are, for example, text documents, tables with figures or drawings. For the proper functioning of features as tools it is of major importance that the skills to use them are easily developed or do already exist with the user. For that, aspects like direct feedback and consistency in procedures are crucial. Therefore, simulations for evaluating the tools' aspects of an interface feature should work as fast as the real design would work and should be accurate in their procedural aspects. Highly interactive simulations are usually required. This may often require considerable programming efforts. In such cases, the only practical approach in a 'concept and sketches' phase is often to evaluate features in bits and pieces and to use independent simulations for different features.

Media

Media such as newspapers, television, etc., serve to get people involved in the topics that are communicated through these media. Similarly, user interface design features can be designed to effectively communicate complex task-related information through the interface (for example, database structures, complex drawings, chemical or mechanical processes). For features to properly function as 'media', the communicated information should be presented such that users can readily perceive and interpret it. It should not be necessary for them to put much effort in figuring out the codes that are used to symbolize the information. Therefore, simulations for evaluating design features for a media role should be accurate in the way information is coded. This means that the requirements for such simulations are largely dependent on the coding dimensions that are used in the design. In many cases, mainly static visual codes are used and paper-and-pencil simulations can be used for evaluations. In other cases, dynamic coding of information is used (for example, as direct feedback to tool usage). In such cases, the media role (which is largely passive and presentation-oriented) can be regarded as the presentation counterpart of the (action-oriented) tool's role. Therefore, combining these two roles in one simulation may sometimes be required.

Intelligent assistants

A good assistant understands what you say or intend to say and can actively think and work together with you. Such assistants may even take over tasks from you and self-initiate new tasks. In some cases, parts of a user interface play the role of an intelligent assistant. Elements that play this role should correctly infer user intentions and goals, should act according to those intentions and goals, and should provide users with accurate expectations, minimizing false hopes. At the same time, users should still keep the feeling that they are in control. For simulations, it is required that the behaviour of the simulated assistant is accurate in terms of the aspects mentioned above. One can think of applying a Wizard of Oz approach for simulating the assistant's behaviour or of trying to program the behaviour of small and independent parts of the assistant's behaviour.

APPLICATION EXAMPLE

Suppose a designer is in the 'concept and sketches' phase of designing a new paint program. The program is meant for incidental use, as well as for routine use in people's daily work. The designer has decided that the program should be very similar to existing paint programs like MacPaint® on Apple Macintosh®, but thinks of adding a few extra features to it. One of the new features is a function to simulate the use of a marker pen. An icon representing the marker function is added to a tools palette that looks similar to that of MacPaint®. The tip of a conventional marker pen usually has an elongated form, so the thickness of the drawn line varies with the drawing direction and with rotation of the marker (see Figure 14.1).

On the screen, the marker tip is represented by a cursor with the form of a small line segment, indicating both the place and angle of the marker. The segment can be rotated, using the menu items 'Rotate Clockw.' and 'Rotate Countercl.w.' of the 'Marker' menu. Every time one of these menu items is selected the marker tip rotates 45°. The tip can also be rotated by using keyboard shortcuts. Different types of markers can be selected with the other items in the 'Marker' menu, but also by double-clicking the icon that represents the marker tool. Initially, when a user starts drawing, the line that is drawn is represented in outline until the draw action stops and the outline is automatically filled.

In the program's 'concepts-and-sketches' phase, the designer wants to conduct a

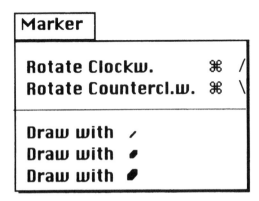

Figure 14.1 The 'Marker' menu.

user evaluation and needs simulations for that. The proposed strategy is applied to find out what is required from the simulations.

Step 1 of the strategy reminds the designer that most of the features of the program are standard and well known to many users. Therefore, the simulation and evaluation efforts will have to concentrate on the 'marker pen' feature. It would not be very efficient to also evaluate the other features of the program; they are fairly standard and it seems better not to change these anyway.

Step 2 tells the designer to analyze the selected feature (the marker function) for the roles it plays in the interaction. In initial use, it is important that users can find out that the marker tool exists, that they can select different markers and that a marker can be rotated. Also, they should be able to find out how to select a marker, what procedure to use for rotating it, and what its general function is. Aspects of the feature that help in doing that, play a 'routing signs' role; these are, for example, the words in the menu, the name of the menu and the visual appearance of the icon. The keyboard shortcuts for rotating the marker tip as well as the double-click option for selecting marker types are intended for use by routine users and play a different role. The designer might want to find out whether rotating by fixed angles is acceptable to experienced users or whether they would prefer a two-step approach in which 'rotation' is selected and the marker tip is subsequently dragged to the desired angle. Such aspects play a 'tools' role in the interaction. Concerning the visual appearance of the drawn line, the designer might want to find out whether the continuous feedback given during the drawing action is acceptable, or whether a filled line is needed as feedback. This feedback aspect plays a 'media' role in the interaction. It should contribute to giving users the feeling that they can observe directly how their actions affect the drawing.

In step 3 the designer decides which roles to concentrate the simulation and evaluation efforts on, and might decide to focus on the 'routing signs' role. One of the possibilities would be to simulate the menu structure, perhaps using menus on Post-It® memos. To investigate the menus for this role it is not necessary that the marker function actually works. If users can select it and get feedback on the effect of the selection, that would be enough for the evaluation. Users could be asked to perform tasks involving the selection of commands and tools in the menu structure by pointing with a finger, while the experimenter manually changes the menus and screen information.

Had the designer decided to investigate the menus for the 'tools' role, then it would have been required that the keyboard shortcuts for rotating the marker tip could actually be used or that double-clicking the marker icon was possible. Obviously, a static simulation with drawings on paper would not do in that case. Instead, one could then think of creating, for example, two separate, interactive simulations: one for evaluating the 'fixed angle rotation' approach and one for evaluating the 'drag–rotation' approach. Subjects in an evaluation could then be asked to reproduce a number of drawings that can be made by just applying the marker tool; drawings that are more or less representative for their usual work. The continuous feedback that is present during the act of drawing would not have to be realistic, as long as its response times are short enough and the feedback is informative enough to create the carefully selected test drawings. However, if the feature is required to be evaluated for the '*media*' role as well, then realistic simulation of the continuous feedback would also be required.

DISCUSSION

The strategy described here is based on findings from research and case studies on the design and evaluation of user interfaces. The requirements are also supported by reports in literature describing case studies, theories or models of human–computer interaction and psychology.

In one case study of the use of paper-based simulations (Vermeeren, 1991) we found that the use of paper simulations for evaluation purposes can indeed lead to valid evaluation outcomes if a program is evaluated for initial learning problems (something that is closely related to the role of routing signs).

In another case study, we designed and evaluated table-making software, and used two different simulation types for investigating the tools aspects of the design (Vermeeren and Kolli, 1993). Psychological principles of skill acquisition, as described by Hammond (1987), helped us in explaining why one of the simulations was deficient, especially for evaluating the role of the tool (as we have seen, for tools the application of skills is a very important aspect).

Maulsby *et al.* (1993) describe the use of a Wizard of Oz approach for simulating an intelligent agent (a very explicit version of an intelligent assistant as we understand it). They find that such a simulation can be used, provided that some specific requirements regarding the Wizard's behaviour are taken into account.

Finally, our strategy bears upon a large number of graduation design projects done at the Faculty of Industrial Design Engineering of Delft University of Technology and for various companies. In these projects a large variety of simulations were used for evaluating user interface designs early in the design process. The strategy has not been applied in practice yet. However, we strongly believe that it can be very helpful in reducing the time needed for creating user interface simulations for user evaluations early in the user interface design process.

ACKNOWLEDGEMENTS

The author would like to thank Mathilde Bekker, Raghu Kolli and Kine Sittig for their fruitful discussions and comments on earlier drafts of this chapter.

REFERENCES

BEKKER, M.M. and VERMEEREN, A.P.O.S. (1993) Developing user interface design tools: An analysis of user interface design practice, in: E.J. Lovesey (Ed.), *Contemporary Ergonomics*, London: Taylor & Francis.

BØDKER, S. (1991) *Through the Interface: A Human Activity Approach to User Interface Design*, Hillsdale, NJ: Erlbaum.

HAMMOND, N. (1987) Principles from the psychology of skill acquisition, in: M.M. Gardiner and B. Christie (Eds.), *Applying Cognitive Psychology to User Interface Design*, Chichester: Wiley.

MAULSBY, D., GREENBERG, S. and MANDER, R. (1993) Prototyping an intelligent agent through Wizard of Oz, in: *Proc. INTERCHI '93*, New York: ACM.

VERMEEREN, A.P.O.S. (1991) Simulation-based evaluation, in: E.J. Lovesey (Ed.), *Contemporary Ergonomics*, London: Taylor & Francis.

VERMEEREN, A.P.O.S. and KOLLI, R. (1993) Feasibility and usefulness of involving users early in the user interface design process: A case study, in: E.J. Lovesey (Ed.), *Contemporary Ergonomics*, London: Taylor & Francis.

VERPLANK, W. (1989) *CHI'89 Tutorial Notes: Graphical Invention for User Interfaces*, San Francisco, CA: ID Two.

PART FIVE

New Evaluation Methods

Feature checklists: a cost-effective method for 'in the field' usability evaluation

E.A. EDGERTON

University of Paisley, Paisley, Scotland, UK

WHAT ARE FEATURE CHECKLISTS?

The feature checklist (FC) is a paper-and-pencil method that gathers information about the features in a system from the user's memory; it is therefore a retrospective measurement. It consists of a list of features of the system under investigation; the intention is that the user will recognize and remember these features which will then act as a cue for asking questions about them.

The usual layout of a FC is a list of features e.g. menu commands, against which are a few columns, each asking a particular question about that feature. An example of a FC and the questions that it might contain is given in Figure 15.1.

The type of information that FCs can elicit will depend on the questions (columns) that the researcher decides to use. The primary aim is to gather information about what features are used and how frequently (usage information). An additional aim is to gather information about users' knowledge of the features in a system, e.g. whether features exist and what they are for (knowledge information). Finally, FCs can gather information about users' attitudes to the features in a system, e.g. whether the user finds the features useful or ever needs them (opinion/attitude information).

FCs can therefore be seen as an alternative to a variety of other measurement methods such as electronic data logs (usage information) and 'thinking aloud' protocols (knowledge information), questionnaires (opinion/attitude information). However, rather than put forward the case that FCs are simply an alternative to other measurement methods, we propose that there is a strong need for FCs in human–computer interaction (HCI) evaluation and that they have many advantages over other methods.

IS THERE A NEED FOR FEATURE CHECKLISTS IN HCI?

Usability testing in industry can occur early in the system development process or it can be done on existing systems that are already 'up and running' (e.g. to produce updated versions). FCs are concerned with the latter of these situations. There are three main reasons why FCs are needed for conducting usability tests of existing systems.

(1) The need for 'real-life' evaluations

In many industrial settings usability testing needs to be conducted in a 'real-life' situation, i.e. the environment where the system exists. This requirement is acknowledged by Bevan and Macleod (1994): 'It is not meaningful to talk simply about the usability of a product, as usability is a function of the context in which the product is used. The characteristics of the content (the user, tasks and environment) may be as important in determining usability as the characteristics of the product itself.' 'Real-life' evaluation has the advantage of avoiding the 'artificiality' of laboratory research, but requires cost-effective methods that can be applied with minimal disruption to the working environment. These requirements exclude many 'traditional' methods.

Electronic data logs are seldom used in real-life evaluations because of the infrequent and unpredictable nature of user behaviour. In order to collect the required information the electronic data log may have to be left on for a long time; this causes software memory problems as well as the problems of translating huge and often irrelevant information from the logs (e.g. mouse positions). Potentially, logging is of great value; however, so far in practice it is poorly matched to investigator needs.

Video and direct observation methods also suffer from similar problems, and also have the unwelcome problem of being obtrusive, i.e. user performance levels may be altered because they are aware that their performance is constantly being monitored. They are therefore not suited to 'real-life' evaluation.

Since FCs are a retrospective method they can be conducted at a time suitable to the user and therefore cause little disruption to working practice.

(2) The need to obtain data from large samples of users

Draper (1985) investigated the behaviour of users of the UNIX operating system. One of the findings of this study was that users typically use only a small, but diverse, personal subset of the many features. The implication of this is that in order to evaluate the system properly large samples of users should be involved to ensure that the data collected are representative of the population as a whole.

However, collecting data from large groups of users can be expensive and requires methods that are cheap to administer. Semi-structured interviews are a valuable but expensive measurement method in terms of researcher and user time; they are therefore not a practical method to use with large samples of users.

Questionnaires can be time-consuming to design and complete depending on the system that is being investigated and the information required; however, they can be used to obtain information from large numbers of users at little extra cost, e.g. by posting them to users. Like questionnaires, FCs can also be used to obtain data from large samples of users. If it can also be established that they are even cheaper to design and complete then they may be a very useful method.

(3) Costs involved

A final reason for using FCs concerns the need to keep costs (in terms of time) to a minimum. The amount of time involved in employing different measurement methods can be broken down into four components:

- *Preparation*, i.e. the time taken by the investigator to prepare the method.
- *Administration*, i.e. the time taken by the investigator to administer the method.
- *Analysis*, i.e. the time taken by the investigator to analyze the data obtained.
- *User*, i.e. the time taken by the user to complete the method.

Table 15.1 compares the FC and the main alternative measurement methods against each of these time constraints.

Electronic data logs require little preparation by the investigator provided the software is available, otherwise they are unlikely to be a feasible method. If they are available the investigator does not need to be present, except to start/stop the log recording (there may be some scope for automating this). Perhaps the greatest amount of time involved in employing this method is that spent by the researcher analyzing the resulting data (discussed earlier). Since electronic data logs are often only practical in laboratory evaluations the cost in time to the user may then be significant (i.e. presence required for the session).

With questionnaires, a significant amount of investigator time is required to design the questionnaire itself; however, they can be administered to large samples of users without the need for the investigator to be present, thus reducing administration time to a minimum. The amount of investigator time required to analyze the data and the time spent by the user to fill in the questionnaire will vary depending on its size and level of detail.

Semi-structured interviews require similar amounts of investigator time in

Table 15.1 Cost (in time) of employing various HCI measurement instruments

Method	Preparation	Administration	Analysis	User
Data logging	low or high	low	high	low or high
Questionnaires	medium/high	low	medium	variable
Semi-structured interviews	medium/high	high	medium	high
Feature checklists	low/medium	low	medium	low/medium

preparing the method and analyzing the data to questionnaires; however, they are often found to be more reliable. A major problem with using semi-structured interviews is that they are costly to employ in terms of both the time spent by the investigator administering them and the time spent by the user completing them.

Feature checklists can be relatively cheap to prepare depending on the system under investigation; they basically involve creating tables listing the features of the system (e.g. icons) and have columns (containing questions) alongside. As with questionnaires the experimenter need not be present and FCs are therefore cheap in terms of administration time. It is likely that the data analysis time and the user involvement time will vary depending on the number of features concerned and the questions asked. However, since the feature checklist only requires the user to tick or cross off answers, the user involvement time should be considerably less than that required for most questionnaires.

ARE FEATURE CHECKLISTS ACCURATE?

If FCs are to be a useful measurement method then the accuracy of the information they provide should be comparable to that of the alternative methods discussed above. One possible problem with FCs is that they are a retrospective measurement and as such their accuracy may be affected by the known unreliability of human memory (e.g. Nickerson and Adams, 1979; Mayes *et al.*, 1988).

Despite the fact that retrospective measurement methods are often suspect, it should be pointed out that in many 'real-life' situations, evaluations are frequently not conducted until after the system has been up and running for a period of time; as a consequence, retrospective methods are often the only option. Although the development of FCs is still at an early stage, preliminary research suggests that the data obtained are accurate and therefore valid; this is especially true for usage information where FCs reliably achieve accuracy scores around 87% (Edgerton, 1992, 1994; Edgerton *et al.*, 1992).

FEATURE CHECKLISTS AND USABILITY TESTING

A key issue in usability testing is not only to detect bugs but also to grade the bugs according to the cost to the user *and* to weight these bugs by frequency of occurrence. Many measurement methods have ignored the latter because it was assumed that user tasks were known (and ignored actual use and practice).

A logical first step is to obtain survey information on tasks (or at least command usage) from as complete a population as possible. FCs are the best way of doing this, and in addition they can detect bugs themselves. By analyzing users' answers to the typical FC questions (see Figure 15.1) it is possible to identify a number of issues that relate to interface bugs. The three most likely issues are:

- *Information flood*, i.e. cases in which users know that the command exists and

File	1 Existed?	2 Used?	3 How often?	4 What for?	5 Need?
New					
Open...					
Close					
Save					
Save As...					
Print Preview...					
Page Setup...					
Print...					
Print Merge...					
Quit					

Column definitions:

Column 1: Existed?   — 'Did you know this command existed?'
Column 2: Used?   — 'Have you ever used this command?'
Column 3: How often?   — 'How often do you use this command (approx.)?'
Column 4: What for?   — 'Do you know what this command does?'
Column 5: Need?   — 'How often do you have any need for this command?'

Answer categories:

For columns 1, 2 and 4, please use the following answer categories:

✓ = 'Yes'   ✗ = 'No'   ? = 'Unsure'

For columns 3 and 5, please use the following answer categories:

0 = never used it/no need at all   3 = once a week
1 = less than once a month   4 = every 2–3 days
2 = once a month   5 = every day

Figure 15.1 Typical FC questions. (The column definitions and answer categories shown above would appear in this format on the front page of the FC. Throughout the rest of the FC they would appear in a summarized format in the header, so that users would not have to keep looking at the front page to remind themselves.)

what it does, yet express little need for using it. If there are many cases of this, the interface may be swamped by features unwanted by the user.

- *Guessability*, i.e. cases in which users know that the command exists but don't know what it does. This may indicate cases where command names are poor at conveying their function to users.

- *Reminding*, i.e. cases in which users know that the command exists, and what it does, yet express that their need to use is greater than their actual usage. This is perhaps a problem of being reminded at the right point in the relevant task.

If FCs are a suitable instrument for detecting the existence of the bugs described, then they may provide system designers with useful information on the appropriateness of command names, the location of icons, etc. However, it is unlikely that major design changes would be based purely on feature checklist information. Instead, it is proposed that after possible bugs have been highlighted, these should then be followed up by other methods, e.g. interviewing a small, representative sample of users.

This represents a crucial aspect of applying FCs; because they have the potential of being a cheap survey instrument they can gather the frequency information (of both usage and some problems) that can focus attention on where the important aspects of a design are, and so direct other instruments that yield better detail but could not be applied across many users, tasks and situations.

APPLYING FEATURE CHECKLISTS IN INDUSTRY

Since FCs are a new measurement method there are few cases where they have been applied in industrial settings. However, one recent pilot study of FCs was part of a project to monitor the use of an electronic document viewing system within a large organization. This system had been installed for three months and the company wished to assess the usage of the system and highlight any major problems that users were experiencing.

The system contained pull-down menus and an icon bar similar to a word-processing application such as WORD 5. One of the main findings that emerged from the study using FCs was that users had not explored (used) many of the features that the system provided despite the fact that they knew what they were for. Based on these findings a number of suggestions were made for increasing usage and understanding of the system.

As far as the FC was concerned, experienced users of the system found the FC much easier and quicker to complete than the questionnaire that was also used. Additionally the FC had a return rate of 62.5%, compared with 51% for the questionnaire.

SUMMARY

The main constraints on using FCs in usability testing are that:

- They should be used to evaluate existing systems.
- They should be used in conjunction with other methods.
- They should be used with experienced users of the system, i.e. users who have had time to explore the functionality of the system.

It is proposed that there is a need for FCs in HCI assessment because they are likely to have the following properties:

- They rely on users' memories and so can draw upon interactions in normal work situations.

- They are a cheap method for the investigator to design and administer (in terms of time).

- They should be a relatively cheap method for the user to complete (in terms of time and effort).

These properties mean that FCs have the following advantages: (i) they are well suited to 'real-life' HCI evaluation research, and (ii) they can be used with large samples of users at little extra cost, and hence can be used to survey entire populations rather than a few individuals of unknown representativeness.

REFERENCES

BEVAN, N. and MACLEOD, M. (1994) Usability measurement in context, *Behaviour and Information Technology*, **13**(1–2): 132–145.

DRAPER, S.W. (1985) The nature of expertise in UNIX, in: B. Shackel (Ed.), *Human–Computer Interaction – INTERACT '84*, pp. 465–471, Amsterdam: North-Holland.

EDGERTON, E. A. (1992) *A Comparison of the Feature Checklist and the Open Response Questionnaire in HCI Evaluation*, Computing Science Research Report, GIST-1993-1, University of Glasgow.

EDGERTON, E.A. (1994) Evaluating feature checklists as a measurement instrument in human–computer interaction, PhD thesis, University of Glasgow (unpublished).

EDGERTON, E. A., LAFFERTY, C. and COOPER, G. (1992) *Visual Realism in Feature Checklist Design: Implications for Validity*, Computing Science Research Report, GIST-1993-2, University of Glasgow.

MAYES, J.T., DRAPER, S.W., McGREGOR, A.M. and OATLEY, K. (1988) Information flow in a user interface: The effect of experience and content on the recall of MacWrite screens, *People and Computers IV*, pp. 275–289, Cambridge: Cambridge University Press.

NICKERSON, R.S. and ADAMS, M.J. (1979) Long-term memory for a common object, *Cognitive Psychology*, **11**: 287–307.

CHAPTER SIXTEEN

Co-discovery exploration: an informal method for the iterative design of consumer products

J.A.M. (HANS) KEMP and T. VAN GELDEREN

Philips Corporate Design, Eindhoven, The Netherlands

INTRODUCTION: THE CONTINUUM OF USE

As consumers, we are used to evaluating the products that surround us, although this is certainly not a simple rational process. In this chapter we will argue that if you want to know how consumers evaluate a particular product, you have to look into the entire evaluation process, starting at the beginning, that leads to the actual purchase of the product. A common model of this process is the problem-solving model (Engel *et al.*, 1990). In its most extended form, the process entails an evaluation of alternatives before purchase. This evaluation is made on the basis of information gathered after the initial need recognition. After the purchase, the consumer evaluates whether his or her choice was the correct one; the outcome of this can be used the next time a similar product is bought. During this evaluation of alternatives, it is not just the extent to which the product meets expectations in terms of efficiency and effectiveness that determines user satisfaction, but the other benefits of buying the product also play an important role (Engel *et al.*, 1990). This is described as the motivation process of consumers (Figure 16.1).

The evaluation is made by comparing the needs with the expected benefits of alternative products. These benefits can have either a utilitarian or a hedonic character. The utilitarian benefits of a product are determined by what are known as the objective product attributes. The hedonic or experiential benefits are determined by the subjective or emotional attributes of the product. This twosome is also referred to as the functional versus the psychosocial aspects of product use, or the 'think/feel' dimension (Snelders *et al.*, 1993). Studies of the user interface of a product should therefore not be restricted to the functional, objective product attributes, whose relative importance will change with time. The way a product is

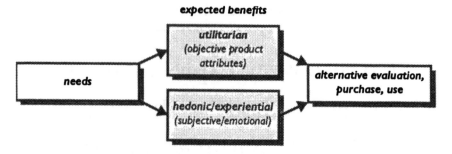

Figure 16.1 Motivation process of consumers (adapted from Engel *et al.*, 1990).

perceived at the very first encounter differs considerably from the way it is perceived after several months of daily use. In other words, the product's user interface requirements will gradually change along the so-called 'continuum of use'. As the authors see it, the continuum of use can be divided into three phases: the first impressions, initial use, and habitual use phases.

The first impressions phase

This phase starts when someone becomes aware of a product for the first time, e.g. by seeing it in an advertisement or in a shop window, or perhaps because a friend demonstrates his or her latest acquisition. The first impression of the product will create certain expectations as to what is offered (e.g. functionality, ease of use or 'quality image'). Most of the issues during this phase deal with subjective and emotional product attributes:

(a) Labelling: does it speak for itself?
(b) Design: does it clearly radiate the benefits of the product?
(c) Overall appearance: what general impression does the product give?

The initial use phase

The initial use phase consists of the user's initial interactions with the product before he or she becomes familiar with using it. During this phase, the user is still learning. A very likely place to start is the shop, where the potential purchaser perhaps tries to perform some self-imposed basic tasks involving the product, or immediately after the purchase, when he or she has installed the product at home and is starting to use it. Typical issues during the initial use phase are:

(a) Feedback: is sufficient relevant information provided?
(b) Self-explanation: is it immediately clear how to operate important functions?
(c) Conventions: is it relevant to what users are accustomed to?

The habitual use phase

We define habitual use as the user's behaviour as it initially evolves but which remains more or less stable after he or she has used the product for some time. In this phase it is crucial that the product has met the user's expectations during the previous phases; it may mean the difference between a satisfied and a disappointed customer. The resultant positive or negative image of the product will certainly be taken into account the next time a similar product is bought. The emphasis during this phase is typically on functional product attributes. Characteristic issues during the habitual use phase are:

(a) Efficiency: can the product be operated with minimum effort?
(b) Convenience: is it still easy to use even after a period of time?
(c) Functionality: does it meet the user's needs?

These three phases can be seen as a continuum: though in each of these phases the user places a different emphasis on the product's requirements, and there is no clear borderline between them.

EVALUATING USABILITY

Assessment of usability in the three phases of the continuum of use has been a topic in user interface research for many years. The issues of the habitual use phase in particular and to a lesser extent the initial use phase (when there is a focus on functional product attributes) can be addressed by several evaluation methods. These methods range from summative methods such as heuristic evaluation (Nielsen, 1992), or practical evaluation methods (Johnson *et al.*, 1989), to more formative evaluations such as cognitive walk-throughs (Polson *et al.*, 1992) and controlled experiments. The research questions of these methods basically address specific utilitarian topics. However, the issues that typically relate to the first impressions phase and, to a lesser extent, the initial use phase (as described here) have so far received only limited attention from usability evaluators. In our opinion, we need an evaluation method that addresses the first impressions and initial use phases in particular, i.e. that reveal the hedonic/experiential aspects of product use.

   Another major requirement of such a method is that it should be possible to communicate the results convincingly to people who are not human factors specialists. It is important for all those concerned to participate in the analysis of the results, in order to stimulate creativity and generate wide support for the conclusions. In this respect, from previous experience we have learned that to issue a report containing the conclusions of some well-controlled experiment is not enough. The rest of this chapter describes a method that attempts to address these issues.

CO-DISCOVERY EXPLORATION

The method we propose in this chapter addresses the first two phases of the usability continuum. The idea is to capture the user's reactions and concepts when he or she is confronted with certain products for the first time (such as when window shopping), and their experiences during the initial use phase (see above). This implies that we want our subjects to interact with the products with as few instructions as possible. To get the subjects to reveal what they think of and how they feel about the products, we invite pairs of subjects and ask them to perform tasks and answer questions together. We then videotape the dialogues and interactions with the products.

The use of pairs of subjects is not a new approach. O'Malley *et al.* (1984) used this technique to elicit verbal protocols in problem-solving analysis, and Hackman and Biers (1992) found that pairs of users make statements which have a higher value for designers than single users thinking aloud for the benefit of an experimenter. Very similar to our approach is co-discovery learning, described by Kennedy (1989), in which two subjects are involved in usability testing. This approach focuses on the utilitarian aspects of the interface (time taken to complete a task, number of errors, etc.), although co-discovery exploration has been designed to investigate hedonic/experiential aspects.

To get the pairs of subjects we invite individuals and ask them to bring along someone he or she knows well. By using pairs of individuals who are familiar with each other, we are able to ensure that the subjects concentrate on the products rather than on each other, and it also helps to ease the tension. A drawback of this approach is, of course, that we have limited control over the participants in the investigation, but nevertheless, we think that this is outweighed by the benefits. It is important that the invited subjects are told as little as possible about what is going to happen because otherwise they may get an expert to accompany them.

The overall structure of a co-discovery session is shown in Figure 16.2. During an introduction (which is not given in the room where the actual session will take place) the subjects are told about the aims and procedure of the session, and given information about the video cameras and microphones in the room that will be used to record the session. Information about the specific product(s) is avoided as far as possible, so that when subjects enter the room, their reactions will be genuinely spontaneous. The experimenter and others who may be present (e.g. product

Figure 16.2 Structure of a co-discovery session.

developers) observe the session from another room and are not visible to the subjects.

The first part of the session comprises questions and/or tasks that encourage the subjects to explore the products. In many cases there are several products in the room and subjects are asked to answer sequentially (the same) questions on each of them. Typical questions include

- What do you think you can do with this apparatus?
- Can you explain what all the knobs and buttons are for?
- Would you buy this product?

Specific features may be highlighted. By presenting a wide range of stimuli (different kinds of products) the experimenters hope to ensure that the subjects' reactions will exceed the level of 'I like the yellow button better than the red one' (Kemp and Verheij van Wijk, 1992). The subjects are asked to perform a number of tasks that will provide information on the self-explanatory features of the products during the initial use phase. Typically, these tasks are related to the basic functionality of the products. In many cases, these tasks will have already been performed during the exploration part of the session (Kemp et al., 1993). The session concludes with a final discussion, which has several objectives: first, any ambiguous behaviour or remarks during the session can be clarified, and questions raised by the subjects can be answered. Tasks they failed to accomplish also have to be explained.

A frequently complicating factor is that the experimenters may not know exactly what they are looking for, so that preferably the analysis should begin during the actual sessions. This will allow specific details or contradictory statements to be checked with the subjects. Further analysis, combining the video data with notes taken during the sessions and possible logging files listing all user-driven events, can take place later. Only during the session will the main usability problems become apparent (see also Kemp et al., 1993), and what possible solutions can be provided to avoid them. The video material should preferably be analyzed by a wide assortment of people involved with the product, for a number of reasons. First, we feel that the best way to transfer the knowledge arising from the investigation is to share the responsibility in building up this knowledge. Developers should therefore be directly involved in the analysis of the sessions. Second, if the aim is to come up with feasible solutions to the usability problems detected, it is beneficial to involve the persons (partly) responsible for them. Furthermore, different viewpoints will uncover more flaws (not just for the ergonomist, but also for commercial and technical people). Software developers in particular tend to detect events that others may have missed.

Three basic questions need to be addressed during the analysis: (1) what are the major flaws in the current product(s)?, (2) what requirements do users demand from these products?, and (3) how can we design a product that fulfils these requirements better than current products?

So far, two techniques have been used for this analysis. One involves groups of people in unedited discussion sessions; if properly directed, these groups will reach a consensus. The data can therefore be presented as objectively as possible. The main drawback of this technique is (of course) the amount of time involved. In general, the time spent analyzing video data increases significantly as more people participate, owing to prolonged discussions and other factors.

The second technique is group analysis of a compilation of video clips. This means that many issues can be covered within a limited amount of time, although inevitably a selection has to be made in advance. This may sometimes be misleading because video data are readily perceived as being 'objective', while the selection of clips is a subjective interpretation of what has happened during the sessions.

THE NUREMBERG CO-DISCOVERY SESSIONS: A CASE STUDY

Aims.   The aims of the sessions were, first, to make a general assessment of the usability problems of current cellular telephones to enable us to formulate a requirements specification for future user interfaces; and second, to evaluate the PR700/Europorty car phone, then (July 1992) under development at Philips Kommunikations Industrie (PKI), Nuremberg.

Setup.   An overview of the setup is shown in Figure 16.3. Two rooms were available: one was used as the session room where the telephones were displayed, and the other (small) room contained the video-recording gear. Cameras and microphones were kept out of sight of the subjects as far as possible.

Design.   The subjects were recruited from the Nuremberg area. They had various backgrounds (students, managers, housewives, pensioners) and belonged to

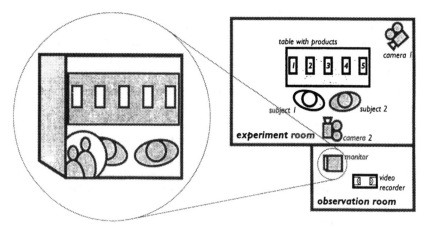

Figure 16.3 Setup of the investigation.

different age categories (ranging from 20 to 70). In all, we had nine pairs of subjects (Virzi, 1992). For each of the five telephones involved in this investigation, the subjects were asked to perform the following tasks:

- Explore the devices: The subjects were asked to give their opinions on every product, without any instructions or manuals, and then to describe their general impression of the devices (which one they liked most, etc.).

- Look for a simple function: The subjects were asked to look for the charge counter in each device. This involved finding their way through part of the functionality and some basic concepts of the user interfaces were evaluated.

- Create a new entry in the directory: Next, the subjects were asked to enter a name and number (of someone they rang frequently) in the telephone directory.

- Answer a call: While they were entering the name into one telephone, we rang the telephone they were using.

- Make a phone call: As a final task, subjects were asked to call the person they had previously entered in the directory.

Final discussion. After the last task had been completed, the experimenter joined them and a somewhat informal discussion took place.

Analysis. The nine sessions resulted in approximately 25 hours of video material, which was edited down to a compilation tape of approximately 50 minutes. The main criteria for selecting the clips were: Is it a representative even:? Is it a remarkable event? Is the image and sound quality good enough?

This compilation tape was then used as the basis for a workshop held in Nuremberg, at which major flaws of the products were recognized as such, and goals were set for future developments. These results were condensed, together with our observations, into a user interface requirements specification for (hand-held) mobile telephones, which is recorded in Kemp *et al.* (1993).

CONCLUSIONS

We believe that the Nuremberg co-discovery sessions have shown that co-discovery exploring is a valuable addition to existing evaluation techniques. It has given us an insight into the user's first impressions of products, as well as the initial usability flaws and merits. It has proven to be both a useful evaluation tool for existing consumer electronics products and a powerful tool for generating new design concepts and redesign. In our opinion, its other major advantages are:

- The subjects are in a relaxed state of mind during the sessions, and on the whole, seem to sincerely enjoy participating.

- The method can improve understanding of how subjects perceive and experience specific products during the first two phases of the continuum of use. Both utilitarian and emotional aspects of the product can also be evaluated.

- The video data produced are relatively easy for everyone to understand. We found it easy to communicate our findings to non-experts, such as product managers and software developers.

At the same time, we encountered several drawbacks of the method. It does not provide a quantitative evaluation of a product; so far, it has proved difficult to produce such 'solid data'. In fact, a proper validation of the method in comparison with other techniques has not yet been done. Also, the process of analyzing the data in such a way that many design solutions are generated is not at all straightforward. This still requires a lot of creativity and skill.

REFERENCES

ENGEL, J.F., BLACKWELL, R.D. and MINIARD, P.W. (1990) *Consumer Behavior*, 6th edn, Hinsdale, IL: Dryden Press.

HACKMAN, G.S. and BIERS, D.W. (1992) Team usability testing: Are two heads better than one? *Proc. 36th Annual Meeting of the Human Factors Society*, pp. 1205–1209.

JOHNSON, G.I., CLEGG, C.W. and RAVDEN, S.J. (1989) Towards a practical method of user interface evaluation, *Applied Ergonomics*, **20**(4), 255–260.

KEMP, J.A.M. and VERHEIJ VAN WIJK, L. (1992) Bingo co-discovery test and requirements specification, *IPO Rapport*, 884.

KEMP, J.A.M., VAN NES, F.L. and TANG, H.K. (1993) MMI requirements specification for handheld mobile telephony, *IPO Rapport*, 897.

KENNEDY, S. (1989) Using video in the BNR usability lab, *ACM SIGCHI Bulletin*, **21**(2): 92–95.

NIELSEN, J. (1992) Finding usability problems through heuristic evaluation, *Proc. ACM CHI '92*, pp. 373–380.

O'MALLEY, C.E., DRAPER, S. and RILEY, M.S. (1984) Constructive interaction: A method for studying human–computer interaction, *Proc. IFIP INTERACT'84, 1st Int. Conf. on Human–Computer Interaction*, London, 4–7 September, pp. 269–274.

POLSON, P.G., LEWIS, C.H., RIEMAN, J. and WHARTON, C. (1992) Cognitive walkthroughs: A method for theory based evaluation of user interfaces, *Int. Journal of Man–Machine Studies*, **36**: 741–773.

SNELDERS, H.M.J.J., SCHOORMANS, J.P.L. and DE BONT, C.J.P.M. (1993) Consumer–product interaction and the validity of conjoint measurement: The relevance of the feel/think dimension, *European Advances in Consumer Research*, **1**: 142–147.

VIRZI, R.A. (1992) Refining the test phase of usability evaluation: How many subjects is enough? *Human Factors*, **34**(4): 457–468.

CHAPTER SEVENTEEN

Private camera conversation: a new method for eliciting user responses

GOVERT DE VRIES, MARK HARTEVELT and RON OOSTERHOLT

Philips Corporate Design, 5600 MD Eindhoven, The Netherlands

INTRODUCTION

A human factors aim is to develop products that fit optimally into users' daily lives, and will become an obvious extension of their actions. Thus, the users' behaviour and requirements during daily life in relation to product use need to be investigated. The conclusions resulting from such investigations are fed into the early stages of the product development process. Early implementation of information about product use enables product developers to include findings in iterative evaluations, leading to the most effective and efficient development process.

Human factors specialists and market researchers have a number of methods at their disposal, such as user observation, interview, questionnaires, focus groups, logging actual use, etc. In recent years some new methods for product evaluation have been developed such as heuristic evaluation (Nielsen and Molich, 1990) and cognitive walk-throughs (Lewis *et al.*, 1990). However, with the successful application of expert-based methods there is a danger that the involvement of real users will decline. Especially in industry, where time is limited and resources are scarce, the involvement of users is often omitted.

Investigations of the use of products, the context of product use and user requirements demand a delicate approach to users in order to reveal personal experiences. Often this is a complex and time-consuming process, so that the selection of the most suitable research method and its proper execution are of paramount importance.

The private camera conversation method offers an opportunity to acquire qualitative and personal information about product use in an effective and efficient way, while entertaining the participants at the same time. The method is based on a Dutch television programme *Achterwerk in de kast,* in which youngsters talk to the camera about a wide range of subjects, such as what happened at school, who do they like or dislike or what television programmes do they enjoy. The youngsters are not interviewed but are asked to talk to the camera and tell their story. The youngsters themselves decide when they want to start and when to stop.

In the private camera conversation method there is no person to talk to and no questions are asked. An expected benefit of this method was that users would not be restricted in their responses by questions, and that they would be less inclined to give only socially acceptable answers. This could lead to more personal, fruitful creative and valid answers. Furthermore, the method would enable us to recruit participants from large public events where people would be willing to participate.

Three case studies were executed in the framework of a research project to identify new methods for eliciting user requirements. The aim of these case studies was to investigate application opportunities and requirements for optimization of the method. Would it bring some of the advantages that were envisioned?

THE METHOD

The three case studies were conducted in different settings, using participants with different characteristics and who were addressed different questions. Using the method with different sorts of users in different contexts and environments provided information about the scope of application. In total, 128 people participated.

Figure 17.1 Illustration of the video cabin.

Common procedures in all three case studies were the following:

1 Visitors at a public event (e.g. an exhibition) were randomly asked to participate in the private camera conversation and were offered the chance of winning a portable CD player.
2 Before entering the video cabin, participants were given instructions by the experimenter.
3 Participants were seated on a bench or in chairs which fixed their position to the camera. The camera and microphone recorded their behaviour (see Figure 17.1).
4 Participants were placed in a private space. They were asked to talk about their ideas, opinions and thoughts on a certain subject to the camera, which was running continuously.
5 The participants themselves decided when the session would start and end. They started the session by opening a curtain so that the camera could see them, and ended it by closing the curtain.

THE CASE STUDIES

Brief descriptions of the aims and the contexts of the three case studies are given in the following.

Case study 1: The Evoluon Exhibition Centre

In October 1993 more than 30,000 people visited the Philips Corporate Design stand at the Evoluon show. The stand provided visitors with information about the work carried out by Corporate Design. The intention was to evaluate the private camera conversation method in a real-life setting.

People visited the Evoluon show together with friends or family. Visitors were invited to participate in the private camera conversation in groups of two or more, and were given a choice of two assignments: to talk about their own VCRs (at home) or to explore a (supplied) portable CD player. It was hoped that these assignments would provide useful information for Philips.

Little is known about the social context in which VCRs are used. Questions such as which family member helps whom, what functions are used by what family member, which functions are regarded as unnecessary and why, how often the VCR is used and by whom, etc., are important for discovering user requirements. The assignment to explore the portable CD player was chosen to determine its learnability and the users' first impressions. Do our customers understand the functionality of the CD player; do they like the design; what aspects will they comment on when not asked specific questions? It could be that certain product aspects that we thought were important are in fact not important at all.

Figure 17.2   Picture from the setting of the first case study.

Case study 2: The Evoluon Exhibition Centre

In January 1994 another show was held at the Evoluon, to which Philips employees and retired Philips employees were invited. This study allowed us to evaluate the private camera conversation method with a different audience – the elderly. The participants were invited to talk about their alarm clocks: What do they like and dislike about them, how could they be improved, how often do they change the alarm time?

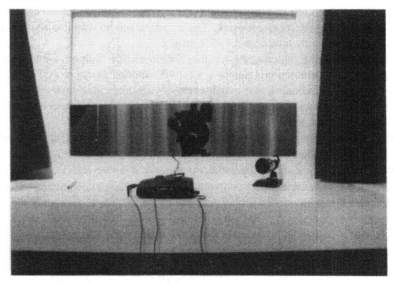

Figure 17.3   Camera setup as seen from the participant's position.

Figure 17.4 Subject setup as seen from behind the camera.

Case study 3: A high school

In this case study the private camera conversation method was used at a high school which had invited Philips to contribute to a school event. The audience consisted of boys and girls 12–18 years old. In this study the aim was to acquire knowledge about what product characteristics enhance an affective relationship between a product and its user. Utilitarian product characteristics (e.g. 'handy', 'effective') can be responsible for this affective relationship, but often other less tangible product characteristics also play an important role. Participants were asked to think of an object they liked very much, an object they were fond of and would not like to lose, and to talk about it as much as possible.

FINDINGS

Spontaneous and informative reactions

Traditional methods are sometimes perceived as boring and time-consuming for participants, but with the private camera conversation method, a context can be created that is very inviting and stimulating. By giving participants limited preparation time, their answers were often spontaneous and informative.

For example, a mother and daughter were asked to talk about how they use their video recorder. The following is a transcript of what they said:

Christie: Ok, hi my name is Christie M.
Janice: And my name is Janice M.
Christie: And we are here to tell you about how we feel about our VCRs. I'll start

since I'm the youngest and most aggressive, ha ha. VCRs!! Mmm VCRs. I use my VCR, which is borrowed, I use it but I don't own it. I use it to record shows late at night when I go to bed and miss. I never have been able to operate the timer. It is too difficult but my husband is able to do it. The biggest reason why we got it was that we had an echo scan and we wanted to be able to play that tape. But overall I wouldn't say that it is a really necessary product in my home since I normally watch what I want to watch when I want to watch it.

Janice: The new ones however, that we should be getting one ooh, seem much more complicated than is necessary and I do think that people designing VCRs should make them simpler to operate.

Christie: Do you have a library of videos, or do you keep retaping the same tape?

Janice: Mmm, I just have a few things in the library but mostly I retape the same tape.

Christie: That's the same with us. No library because most TV shows you don't want to treasure for a long time. So we have three tapes that we keep taping over and over again.

End of session, participants close the curtains.

This excerpt illustrates the participants' behaviour in terms of VCR use and gives an idea of their attitudes to the VCR's interaction design.

Useful silences

During the sessions participants often came up with fresh, sometimes revealing thoughts and ideas just after there had been a long silence. Therefore, in the briefing participants should be asked to try to elaborate on the subject as much as possible and it should be stressed to them that silences are normal and occur regularly. The fact that after a silence additional interesting information was produced shows the benefit of the private camera conversation method; in traditional interview situations interviewees often feel that silences are painful and so try to avoid them.

How to address participants

The way in which participants are approached has an important influence on their willingness to respond. Some people who told very useful and personal stories refused to tell the same story on camera. The stories were rather private and intimate. A good introduction describing the aims of the session is important in putting the participants at ease. The location of the cabin is also important; it should not be too noisy, nor close to activities that do not match the character of the private camera conversation sessions (e.g. building entrance).

Use of questionnaires

In some sessions very brief questionnaires were used to prompt the participants.

The usefulness of the questionnaires varied with the participants; some seemed to be prompted to come up with a lot of extra and very relevant information, whereas others seemed to be restricted in their creativity by the questionnaire and slipped into a question and answer situation.

Use of more than one participant

Most sessions in the case studies were run with two or more participants rather than only one. The company of a familiar person inside the booth appeared to increase the ease with which participants expressed their thoughts.

Differences between adults and children

Youngsters were happy to talk straight to the camera, whereas adults preferred to talk to and interview each other. The reason for this is unclear. It could be that adults take the assignment 'talk to the camera' less literally than children do. It could also be that youngsters feel more comfortable talking to the camera than adults do. The exact reason for this difference should be investigated, since it can have consequences for an optimal application of the method. For example, certain questions might be more appropriate for youngsters, and others more appropriate for adults.

Group dynamics

In the first case study, families were invited to talk about how they used their VCRs in order to observe group dynamics. The roles that people played seemed very natural. Participants mentioned their dependence on others to operate some of the VCR's functions. The discussions about what part of the functionality was understood and used by whom, were very lively and natural.

Session time

The majority of participants took 5–10 minutes to tell their stories to the camera. This was longer than expected; many people felt at ease when participating and had more to say than they had envisioned beforehand.

Camera control

Participants were sometimes unsure whether the camera was running or not, despite the fact that they had been told that it was running continuously. Although opening and closing the curtains in the video cabin proved to be a good way to start and end the sessions, participants were still in doubt whether the video was running. A red light which lights up as long as the camera is running could provide a solution.

DISCUSSION

Find the right application areas

In general, the methods developed for specific human factors research can be used in various applications. For example, the method of interviewing people can be used for setting criteria, for comparing two different user interfaces, or for determining whether an interface is suitable for a particular group of users (Jordan, 1993).

In the case studies using the private camera conversation method, the research questions addressed the social context of VCR use, the learnability of a portable CD player, pleasure in products, and the use of clock radios. More systematic testing of the method is needed to learn what types of questions can be asked of what people, and in what contexts.

Number of participants in the video cabin

For many participants, the presence of a friend or relative seemed to help to reduce the tension and 'stage fright'. People enjoy participating in pairs or in small groups since they can react and respond to each other. However, the most creative and useful thoughts and ideas were often expressed after silences, which are much more likely to occur when only one participant is in the cabin. How to benefit from both situations remains to be explored.

Prompting users

As mentioned before, some participants seemed to be restricted in their creativity and richness of response when a questionnaire was used to prompt them. An idea to overcome this problem is to put a questionnaire in an envelope and to ask participants to open it only if they can think of nothing more to say.

User involvement in protocol analysis

In the projects carried out so far, human factors specialists evaluated the data obtained with the private camera conversation method. These data could be validated and enriched by inviting the participants to look at their own video recordings and to respond to them.

CONCLUSIONS

The private camera conversation method is a very useful means for obtaining feedback from users about their product use. It can provide qualitatively rich and often personal information on various subjects such as the social context in which products are used and determinants of the user's affective relation with a product.

The private camera conversation method is efficient; recruitment, briefing and monitoring the session take very little time, and it can be used easily at public events such as exhibitions. It is interesting for potential participants and can provide a strong company visibility at the same time. Application areas and possibilities for optimization should be further investigated.

ACKNOWLEDGEMENTS

The authors would like to thank Patrick Jordan, Ian McClelland, Irene McWilliam and Lisa Cherian for support and advice.

REFERENCES

JORDAN, P.W. (1993) Methods for user interface performance measurement, in: E.J. Lovesey (Ed.), *Contemporary Ergonomics*, pp. 451–460, London: Taylor & Francis.

LEWIS, C., POLSON, P., WHARTON, C. and RIEMAN, J. (1990) Testing a walkthrough methodology for theory-based design of walk-up-and-use interface, in: J.C. Chew and J. Whiteside (Eds), *Proc. CHI'90* (Seattle), ACM, pp. 235–242, Reading, MA: Addison-Wesley.

NIELSEN, J. and MOLICH, R. (1990) Heuristic evaluations of user interfaces, in: J.C. Chew and J. Whiteside (Eds), *Proc. CHI'90*, (Seattle), ACM, pp. 249–256, Reading, MA: Addison-Wesley.

CHAPTER EIGHTEEN

Repertory grid theory and its application to product evaluation

CHRIS BABER

University of Birmingham, Birmingham, UK

INTRODUCTION

There has been sporadic interest in repertory grid theory among ergonomists. In recent years, repertory grid analysis has been employed in the evaluation of text types (Dillon and McKnight, 1990) and in evaluating collaboration (Schuler *et al.*, 1990). Sinclair (1990) suggests that repertory grids can be useful in areas of product evaluation.

The theory and its methodology have also been profitably applied to study of consumer behaviour. It is possible, for instance, to use personal constructs derived from repertory grids in considering the 'cognitive models' of individuals with varying degrees of experience and expertise. In one study, first-year undergraduate students provided fewer constructs, concerning the retail environment of a large city, than would be expected of people with more experience of the shops. This suggested that the shopping habits of the students would feature less discrimination and, hence, choice, than those of people who had lived in the area for a longer period of time. One can conceive easily of extensions of this approach to consider the constructs of 'novice' and 'expert' computer users.

Kelly (1955) proposed that people act on the basis of specific hypotheses, or expectations, concerning the functioning of their environment, i.e. people are assumed to be 'scientists', developing hypotheses concerning the best course of action to take in a given situation. It is, perhaps, not too much of an extension of the original proposal to relate these notions to the term 'mental model', which has become fashionable within the ergonomics community. A 'mental model' could be viewed as collection of 'constructs' related to a particular product. By this I mean that people can often be seen to use products on a 'hypothesis and test' approach. The hypothesis formed for a specific action is derived from the available information in the product and its current state. Previous work has used this

157

conception to develop a methodology for identifying likely points at which users will make errors in their use of the product (see Stanton and Baber, Chapter 24). An extension of this previous work would be to include some notion of the type of information upon which the users' hypotheses will be based. At a general level, it is possible that the information can be related to users' constructs.

For Kelly's theory, the notion of constructs assumed that people would focus on stable aspects of an environment, which could be expected in future experience of that environment. Furthermore, he proposed that constructs would exist in terms of polar opposites, e.g. 'good'/'bad', but that the oppositions between a construct and its opposite, or 'contrast' term, would be defined by the individual rather than from any external, objective source. Thus, the theory was labelled 'personal construct' theory.

A repertory grid illustrates the relationship between an individual's construct/ contrast sets and a range of items (in this case, consumer products). The initial phase of collecting a subject's constructs is based on the method of 'triads'. From the notion of constructs existing as ideographically defined, polar opposites, one can assume that, for any three items, two can be assumed to be similar on one dimension and different from the third, i.e. to take the example from Sinclair (1990), two telephones could be defined as similar to each other and different from a third, in terms of a 'modern' versus an 'old-fashioned' appearance.

Eliciting triads is thus simply a matter of presenting groups of three items to subjects, and asking them to define a pair and to state how this pair differs from the third item. Obviously, some care needs to be taken at this juncture to ensure that the presentation of items will not unduly influence the subject, i.e. it is important to make sure that the items come from the same product group to enable comparison of like with like. Once a construct has been elicited, the subject can then be asked to state whether other items in the set represent positive and negative instances of that construct. The process is then repeated with another group of three items.

From this brief introduction, it is clear that the method is designed to be used on an individual basis. For an ergonomist, this would suggest that the method would be most useful either for products which have a small number of potential users, or for a product which is still at the concept stage. Having said this, I would agree with Sinclair (1990), that repertory grid analysis could be used in a wide range of projects. A major aim of the technique is to provide the analyst with a vocabulary, based on the respondent's own terms, and to illustrate the inter-relationship between these terms.

Texts which explain how to conduct repertory grid analysis often turn to forms of factor analysis, which seem to extend the original theory on which repertory grids are based. While there is an interest in developing and improving computer packages to assist in the elicitation and analysis of repertory grids, the personal constructs elicited can only be meaningfully interpreted ideographically, i.e. with reference to the individual's 'belief system'. Furthermore, it is difficult to determine how the information derived from the constructs can be scored in any meaningful manner, or to define the distribution of the information, in statistical terms. While the method is flexible enough to deal with a broad range of problems, the complexity of data can lead to emphasis on involved analysis which could obscure errors in the analysts' reasoning. In Kelly's original conception, manual analysis would allow the analyst to 'handle' the data and spot problems during analysis; the use of factor analysis forces the analyst to part from the data until a 'result' is presented. This chapter presents an approach to the analysis of repertory grids which marks a return to Kelly's proposals. It is based on the work of Coshall (1991) in demographics, and is applied to consumer ergonomics. The approach will be illustrated with a worked example concerning an individual's responses to eight commercially available microwave ovens, and four respondents' ranking of wrist-rests.

DEFINING A REPERTORY GRID

The time required for construct elicitation will depend on several variables, such as the willingness of the subject, the similarity between items, and the number of items used. While one could propose that evaluation be repeated until all possible combinations have been exhausted, it would be more pragmatic to proceed until the subject is unable to offer any new constructs. From this process, it will be possible to define a grid relating all items to the elicited constructs (see Table 18.1).

The next phase is to analyze the constructs in order to ascertain possible relationships between constructs, i.e. to determine whether different elicited constructs may be manifestations of the same underlying personal construct. It is at this point that textbooks tend to shift towards factor analysis approaches; after all, stated in these terms, the problem appears to be one of defining appropriate factoring of the information. In the following example, I will demonstrate a simple, alternative version which, while it may lack the statistical rigour of factor analysis, can offer a quick and easy alternative means of repertory grid analysis. As a coda to this point, it is also worth noting that much of factor analysis relies of the intuitions of the analyst in selecting the appropriate range and rotation of factors.

Example 1

In this example, the factors influencing the decisions of a consumer in selecting a microwave are considered. The subject used was female and in her mid twenties. The eight microwaves used are commercially available, and cover a range of different specifications and manufacturers.

The first phase of the study, the construct elicitation, was performed using photographs of the products, combined with some details from the manufacturers. The subject was presented with groups of three photographs and was asked to provide a construct to define a pair. This phase lasted approximately 25 minutes, and the resulting constructs are presented in Table 18.1. A one (1) indicates an agreement between a construct and an item, and a zero (0) indicates an agreement between the item and its contrast (although there were no instances in this example, there will be cases in which a particular construct/contrast simply does not apply to an item. In this case, the cell is left blank).

Table 18.1 Initial repertory grid and first pass analysis

| \multicolumn{8}{c}{Item number} | | | | | | | | | | |
1	2	3	4	5	6	7	8	Construct	Contrast	Fla
1	1	1	0	0	0	0	0	dials	touch pad	−7
1	1	1	1	0	1	0	1	<800 W	>800 W	4
0	0	0	1	1	1	1	1	clock	no clock	7
1	1	1	1	1	1	0	1	white	black	5
0	0	0	1	1	1	1	1	timer (90 min)	<90 min	7
0	0	0	1	1	1	0	1	memory	no memory	8
0	0	0	0	0	0	0	1	grill	no grill	5
1	1	0	0	1	0	0	1	<5 settings	>5 settings	0
0	0	0	0	1	1	1	1	defrost	no defrost	6
1	0	1	1	1	1	1	0	button (door)	lever (door)	0
1	1	1	1	0	0	0	0	<£130	>£130	−6
0	0	1	1	0	0	0	0	fitted plug	no plug	0
0	0	0	1	1	1	1	1	delay start	no delay	7
1	1	1	1	0	1	0	1	<0.8 ft^3 cap.	>0.8 ft^3	0
7	6	7	10	8	9	5	10			

Once the initial repertory grid has been constructed, analysis can commence. If we remember that we are looking for constructs which may be related, then we can propose that constructs which exhibit similar responses may be related. This proposal is the foundation for the method discussed in this chapter; rather than creating factors and grouping statistically, this approach requires the grouping to be performed manually.

In order to compare patterns of ones and zeros it is necessary to define a 'reference pattern'. The creation of a reference pattern is very simple; we count the number of ones in each column and enter the sum at the bottom of the columns (see Table 18.1). This produces a row of numbers. The next step is to convert the row into a 'reference pattern'. We take the bottom row, containing the sums from the columns, in this case 7 6 7 10 8 9 5 10, and convert the digits into a pattern of ones and zeros. The manner in which this row is split is somewhat arbitrary; we divide the row into two parts, scoring the higher numbers, i.e. assume a cut-off of 7, then for Table 18.1, we would have a reference pattern of 0 0 0 1 1 1 0 1. In this example, the row is split into two equal halves, but in other instances there may be an imbalance between the scored and unscored numbers which needs to be considered by the analyst.

The reference pattern is now compared with the rows for each construct. The procedure now is to count the number of matches between ones and zeros between each row and the reference pattern, out of the number possible, defined by items (in this example eight). The total number of 'hits' is entered in the column labelled 'Fla', to the right of the contrast column in Table 18.1. This in essence is the main data reduction stage.

Table 18.2 Repertory grid after 'reflection'

1	2	3	4	5	6	7	8	Construct	Contrast	F1a
0	0	0	1	1	1	1	1	touch pad	dials	7
1	1	1	1	0	1	0	1	<800 W	>800 W	4
0	0	0	1	1	1	1	1	clock	no clock	7
1	1	1	1	1	1	0	1	white	black	5
0	0	0	1	1	1	1	1	timer (90 min)	<90 min	7
0	0	0	1	1	1	0	1	memory	no memory	8
0	0	0	0	0	0	0	1	grill	no grill	5
1	1	0	0	1	0	0	1	<5 settings	>5 settings	0
0	0	0	0	1	1	1	1	defrost	no defrost	6
1	0	1	1	1	1	1	0	push button	lever	0
0	0	0	0	1	1	1	1	>£130	<£130	6
0	0	1	1	0	0	0	0	fitted plug	no plug	0
0	0	0	1	1	1	1	1	delay start	no delay	7
1	1	1	1	0	1	0	1	<0.8 ft^3 cap.	>0.8 ft^3	0
5	4	5	10	10	11	7	12			
[0	0	0	1	1	1	0	1]			

The reader will notice that there are some negative numbers in column F1a. This occurs because, for some constructs, there were more misses than hits. As a rule of thumb, if there are an equal number of hits and misses enter '0' in the F column for a construct; if there are more hits than misses enter a positive score for the sum of hits; and if there are more misses than hits, enter a negative score for the sum of misses. Again this process is somewhat arbitrary. However, after working through the technique the 'logic' behind this ought to become apparent.

The aim of the analysis at this stage is to ensure that the arrangement of the grid has been optimized, i.e. to eliminate negative scores. In Kelly's (1955) proposals, a negative score resulted from an inversion of the construct, and by turning the construct/contrast around, it is possible to remove the negative score. This process is known as 'reflection', and Table 18.2 shows the analysis after reflection of constructs.

Once again, we split the bottom row into two to form a reference pattern, and compare this with the rows. As there are no negative values in the F column in Table 18.2, we can begin extracting common constructs. This phase makes reference to binomial theorem determining the probability of matches between reference and construct rows occurring by chance. Rather than delve into the relevant mathematics, I have appended a table of binomial distributions (see Appendix). For a given set of items, e.g. 8, there will be a suitable level of matching which can be assumed to be statistically significant. If we assume that selection of the 0.05 significance level will be sufficient to reduce type one error, then simply by reading the Appendix, we can determine whether any of the constructs can be proposed as a group from the initial analysis. From the Appendix, we can see that, for $n = 8$, there are two numbers which provide significance levels of less than 0.05. These are $8 - 0 = 8$ (0.04) and $8 - 1 = 7$ (0.035). This means that constructs which have a score of either 7 or 8 in the F column can be taken as a group.

Thus, from our initial analysis, we can extract the following set of constructs which are assumed to be, in some way, related:

F1 =   touch pad
       digital clock
       > 90 min timer
       memory
       delay start

This technique does not permit the analyst to determine an accurate level of statistical confidence in this grouping. However, as we introduce a level of <0.05 to extract constructs, and as the grouping appears intuitively appealing, I propose that what the technique may lose in its statistical validity is more than compensated for in terms of speed of operation, and construct validity.

Following the definition of an initial grouping, we then remove these constructs from our set and repeat the above procedure. From this analysis, only one construct emerges:

F2 =   lever/push button door operation

We remove this construct and repeat the procedures, until either we factor all the

Table 18.3 Summary of construct groups with appended labels

Pass	No 'reflections'	Constructs	Factor label
1	1	touch pad digital clock > 90 min timer memory delay start	'technical sophistication'
2	0	lever/push button	'door operation'
3	1	power settings fitted plug	'electrics'
4	2	< 800 W defrost <£130	'buying points'
5	0	< 0.8 ft^3 cap.	'size'
6	0	white grill	'appearance'

constructs or we reach a point at which it is not possible to produce significant scores in the F column. In this particular example, it was necessary to produce a further four factors, from four additional passes:

F3 = power settings, plug
F4 = <800 W, defrost, <£130
F5 = < 0.8 ft^3 capacity
F6 = white, grill

In all, the process of analysis, reflection and factoring for this example took approximately 30 minutes. Following this analysis, the construct groups were presented to the subject for comment and she was asked to provide a label for each factor. This, then, produces a summary of the analysis (shown in Table 18.3).

The resulting groupings were felt, by the subject, to be good indications of her consideration of the products used. Thus, while it is not possible to offer confidence levels or correlational data for this information, one can argue that the method can provide a set of factors which reflect subjective opinion to an acceptable level.

Example 2

While the first example illustrated the procedure for defining a person's constructs, this second example looks at a particular application of the technique. The example is taken from a report by Fearnside (1993) on the design and use of six commercially available wrist-rests for typists. The first stage of this part of the work elicited constructs from four respondents. The respondents then used their constructs to rank order the wrist-rests used in the study. Table 18.4 shows the relationship between constructs and wrist-rests. The rests are ordered from left to right in terms of their ranking.

The ranking of wrist-rests in Table 18.4 allows the analyst to consider the relative strengths and weaknesses of the products. For instance, product 2 features

Table 18.4 Wrist-rests rank ordered against constructs

Construct	Product					
Ergonomics	1	2	3	4	5	6
Appearance	5	2	3	1	6	4
Dimensions	6	3	1	4	5	6
Manoeuvrability	6	1	2	4	5	3
Width	2	1	2	4	3	5
Surface	2	5	3	6	4	1
Comfort	2	5	3	6	4	1
Height	2	5	6	3	4	1
Adaptability	3	2	6	1	5	4

in the first three ranks for all constructs, whereas products 5 and 6 seem to have more disparate ranking across constructs. Obviously, without seeing the wrist-rests, these results are not particularly meaningful. However, the point of this example is to demonstrate the progression of repertory grid for comparative product evaluation.

DISCUSSION

This chapter has reported, and illustrated, a technique for conducting repertory grid analysis by hand. It is proposed that the speed and ease with which the technique can be applied make it an ideal candidate for adoption for research and development projects working to tight time constraints. As the technique is explicitly proceduralized, there seems to be a reasonable interjudge agreement on the resulting construct groupings.

Obviously the issue of reliability will require some consideration. However, it is very difficult to determine an appropriate measure of reliability for a repertory grid; it is subject to a range of individual differences. The principal method of determining reliability is to test for intra-subject consistency over repeated testing. Having said this, the principal benefits of the technique are that it does not require knowledge of statistics nor access to a computer, it is quick and relatively robust, and it produces intuitively appealing construct groupings.

The application of this technique would be most appropriate in early stages of product design, perhaps when considering people's notions of desirable design characteristics. However, one can also envisage the approach being useful as a means of defining users' conceptions of usability, perhaps through a comparison of a range of products which users' are encouraged to consider in terms of functionality, ease of use, etc.

APPENDIX

Table A18.1   Table of cumulative binomial distributions

						x					
N	0	1	2	3	4	5	6	7	8	9	10
5	031	188	500	812	969	1					
6	018	109	344	656	891	984	1				
7	008	062	227	500	773	938	992	1			
8	004	035	145	363	637	855	965	996	1		
9	002	020	090	254	500	746	910	980	998	1	
10	001	011	055	172	377	623	828	945	989	999	1

REFERENCES

COSHALL, J.T. (1991) An appropriate method for eliciting personal construct subsystems from repertory grids, *The Psychologist*, **4**: 354–357.

DILLON, A. and McKNIGHT, C. (1990) Towards a classification of text types: A repertory grid approach, *Int. J. Man–Machine Studies*, **33**: 623–636.

FEARNSIDE, P.J. (1993) An evaluation of alternative wrist-rests for VDU work, MSc thesis, Birmingham University School of Manufacturing and Mechanical Engineering (unpublished).

KELLY, G.A. (1955) *The Psychology of Personal Constructs*, New York: Norton.

SCHULER, D., RUSSO, P., BOOSE, J. and BRADSHAW, J. (1990) Using personal construct techniques for collaborative evaluation, *Int. J. Man–Machine Studies*, **33**: 512–536.

SINCLAIR, M.A. (1990) Subjective assessment, in: J.R. Wilson and E.N. Corlett (Eds), *Evaluation of Human Work*, London: Taylor & Francis.

PART SIX

'Off-the-Shelf' Evaluation Methods

CHAPTER NINETEEN

The software usability measurement inventory: background and usage

J. KIRAKOWSKI

Human Factors Research Group, University College, Cork, Ireland

BRIEF HISTORICAL BACKGROUND TO SUMI

The Software Usability Measurement Inventory (SUMI) is the latest development in a series of studies into questionnaire methods of analyzing user reactions which started in 1986. An earlier landmark was the Computer User Satisfaction Inventory (CUSI) (see Kirakowski, 1987; Kirakowski and Corbett, 1988). Both CUSI and SUMI depend on there being a working version of the software system. A group of typical users is needed, who have experience of doing a set of representative tasks with the system in a known and specifiable environment. The users then rate the system using the questionnaire and the results are analyzed in a standardized manner, to yield quantitative, numerical data. The result is the users' view of the quality of use of the system being evaluated.

Attitude scales such as SUMI, CUSI, as well as Shneiderman's QUIS (see Shneiderman, 1992) and the oft-cited questionnaire by Ives *et al.* (1983) measure users' attitudes to specific software systems. We may distinguish between these specific scales, and general scales which instruct users to rate their attitudes to computers and information technology (see, for instance, Igbaria and Parasuraman, 1991). General scales are useful for the insight they give us into society's attitudes towards computers, and the society's attitudes on the impact of computers in that society. They do not enable the evaluator to focus on specific software systems.

We may also distinguish between specific attitude scales which are applicable to a broad range of software systems, and bespoke survey-type questionnaires which request users to consider aspects of a system which are unique to one or possibly a small family of systems. All attitude questionnaires require several iterations of the cycle of item analysis, data gathering, and statistical analysis before they can be used with confidence. In general, bespoke survey-type questionnaires are not reusable enough to warrant this amount of effort and the literature has many examples of such one-off questionnaires, developed on the spur of the moment, and usually characterized by low reliabilities, when these are actually cited.

Thus SUMI, CUSI, and other developments (see also Lewis, 1991; Igbaria and Nachman, 1991; Ramamurthy *et al.*, 1992) seek to occupy the middle ground between sociological-type surveys and bespoke survey-type questionnaires.

CUSI established two subscales of usability, called at the time 'affect' (the degree to which users like the computer system) and 'competence' (the degree to which users feel supported by the computer system). Studies by Wong and Rengger (1990) and Lucey (1991) demonstrated that the Affect scale of CUSI correlated well with the QUIS questionnaire, and a short questionnaire circulated by John Brookes, called SUS (see Chapter 21). The competence scale, however, was marked by low correlations with these other questionnaires and looked at least bimodal. Comparisons with CUSI, as with all the other satisfaction questionnaires of the time, could only be comparative, that is, system X against system Y.

Work on SUMI started in the CEC-supported ESPRIT project 5429, Measuring Usability of Systems in Context, MUSiC (see also Macleod, Chapter 25). One of the work packages entrusted to the Human Factors Research Group (HFRG) within this project was to develop questionnaire methods of assessing usability. The objectives of this work package were first, to examine the CUSI competence scale and to extract further subscales if warranted by the evidence, and second, to achieve an international standardization database for the new questionnaire. Both of these objectives were achieved. The SUMI questionnaire was first published in 1993 and has been widely disseminated since then, both in Europe and in the United States.

THE SUMI QUESTIONNAIRE

SUMI consists of 50 attitude statements, to each of which the user may respond 'agree', 'don't know', or 'disagree'. A three-point response format was adopted because in early versions of the questionnaire many respondents found it difficult to discriminate between five or more shades of opinion. Typical statements from the current version of SUMI are:

1.   This software responds too slowly to inputs.
3.   The instructions and prompts are helpful.
13.   The way that system information is presented is clear and understandable.
22.   I would not like to use this software every day.

SUMI is applicable to any software system which has a display, a keyboard or other data entry device, and a peripheral memory device such as a disk drive. It has also been used successfully for evaluating the client side of client–server applications.

The minimum user sample size needed for an analysis with tolerable precision using SUMI is of the order of 10–12 users, although evaluations have been carried out successfully with smaller sample sizes. However, the generalizability of the SUMI results depends not so much on the sample size itself, but the care with which the context of use of the software has been studied and the design plan has been made. As summarized by Macleod (see Chapter 25), this involves identifying the typical users of the software, the goals which they typically wish to achieve, and the technical, physical and organizational environments in which the work is carried out (for prototype systems, this involves determining the future context of use). The design plan requires an adequate sampling of the context of use.

With regard to the requirement that there be a working version of the product, this does not turn out to be such a serious limitation after all. Most software is created on the basis of improvements or upgrades to a previous version, or in response to a market opportunity created by gaps in competitive products. Usability evaluation can therefore feed into the earliest stages of system specification, as well as enable the setting of usability targets to be achieved by the new system. Many companies now use some kind of rapid prototyping strategy, especially with GUI environments, and the SUMI questionnaire lends itself ideally to this kind of development work as it is short (it takes a maximum of 5 minutes to complete) and does not require a large user sample.

The questionnaire is administered in paper-and-pen format. Some computerized versions exist, but it is important to restandardize the questionnaire for each implementation, as differences between implementations can appear in the response profiles of users. Besides which, most industrial users, at present at any rate, prefer to use the paper-and-pen format. SUMI may be administered and scored by experienced psychometricians, who will usually need training only in the interpretative aspects of the analysis. SUMI administrators with a psychology background will in addition need training in scoring procedures and interpretation, with usually a review of the statistical procedures involved. Other HCI personnel who wish to use SUMI are strongly advised to also take a refresher course, if they have not already covered such issues recently, in evaluation study design. With appropriate training SUMI has been used successfully by software engineers in an industrial setting as well as by more psychometrically aware researchers in academia.

SUMI is scored and interpreted with reference to a standardization database. This is a database of approximately 2000 user profiles, representing a mixed bag of software products that includes office systems, CAD packages, and communications software. This collection is updated at least once a year to reflect changes in market trends and increased user expectations for usable products. A number of multimedia information systems have also been included in the latest version of the database and SUMI is recommended for multimedia courseware development by Davies and Brailsford (1994), as well as for the development of commercial GUI interfaces (see Redmond-Pyle and Moore, 1995).

SUMI OUTPUTS AND THEIR INTERPRETATION

Levels 1 and 2: Global scale and subscales

SUMI can provide results at three levels of analysis. First, there is a Global score, which reflects overall subjective usability. Then there are five usability subscales:

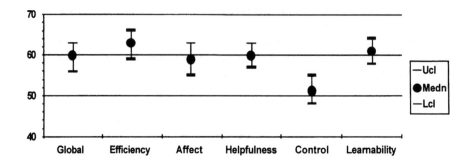

Figure 19.1 Example of a usability profile showing the SUMI subscales, medians, and the confidence intervals around the median.

Affect, Efficiency, Helpfulness, Control and Learnability. The precise meanings of these subscales are given in the SUMI manual, but in general, the Affect subscale measures (as before, in CUSI) the user's general emotional reaction to the software – it may be glossed as Likeability. Efficiency measures the degree to which users feel that the software assists them in their work and is related to the concept of transparency. Helpfulness measures the degree to which the software is self-explanatory, as well as more specific things like the adequacy of help facilities and documentation. The Control dimension measures the extent to which the user feels in control of the software, as opposed to being controlled by the software, when carrying out the task. Finally, Learnability measures the speed and facility with which the user feels that he or she has been able to master the system, or to learn how to use new features when necessary.

An example of a usability profile showing subscale scores is given in Figure 19.1. From this figure it can be seen that the system being evaluated is in general higher than average on all of the SUMI subscales; the users, however, consider that their ability to stay in control of the system is the poorest aspect of their interaction with it. The average for state-of-the-art commercial systems is set by SUMI at a score of 50 on Global and all the other subscales. These dimensions are moderately orthogonal, and at any rate have been empirically discovered using factor analysis. They represent real end users' constructions of the concept of software quality. Statistically independent reliability estimates usually give the Global scale a reliability in excess of 0.92, and four of the five subscales usually show reliabilities in excess of 0.80. The Control subscale yields lower reliabilities, sometimes as low as 0.71 (see, for instance, the reliability estimates from the standardization database of 25 January 1993, cited in the SUMI handbook; Porteous *et al.*, 1993). A recent study on the replication of the SUMI factors has shown, that, on the basis of a totally independent sample of user data, the correlations between the original subscale factors and the replicated ones are of the order of 0.778 to 0.395 (M. Porteous, personal communication). This represents a high degree of stability and indicates that the five SUMI subscales could not have been obtained by chance alone. These figures are satisfactory for industrial use, especially when backed up by the third level of analysis, the Item Consensual Analysis (see below). There is a tendency for each subscale distribution to skew to the upper end of the distribution, perhaps because the standardization database is made up of state-of-the-art commercial software.

When we compare the SUMI subscales with the seven dialogue principles of ISO 9241 part 10 (see ISO 9241, 1992), four of the SUMI subscales have an obvious correspondence, affording a direct measurement of the ISO principles. SUMI 'Helpfulness' and the ISO principle 'Self-descriptiveness' are clearly related, as are SUMI 'Control' and ISO 'Controllability'; SUMI 'Learnability' and ISO 'Suitability for Learning'; and SUMI 'Efficiency' and ISO 'Suitability for the task'.

The ISO principle of 'Conformity with user expectations' may be related to SUMI 'Efficiency' or 'Affect', while the ISO principles of 'Error tolerance' and 'Suitability for individualization' may be related to the SUMI 'Control' or 'Helpfulness' subscales. The degree to which software may be seen to exhibit these last three principles may be more visible to a technical software designer than to an end user. In fact, several individual SUMI items can be isolated and used as measures of the users' reactions to these three ISO principles; however, it is significant that the various factor analyses carried out in the course of the development of SUMI with end user samples did not lead to them being identified as proper subscales.

There is no direct ISO correlate of the SUMI 'Affect' dimension; this is of course to be expected, given that the ISO dialogue principles are specifically presented without reference to situations of use. In this regard, the recommendations in ISO 9241, part 11 (ISO 9241, 1994) with regard to the inclusion of satisfaction ratings in evaluation are to be considered highly significant.

Item Consensual Analysis and profile analysis

The third level of analysis, called Item Consensual Analysis (ICA), is a method of questionnaire analysis specifically developed for the SUMI questionnaire. In Item Consensual Analysis, a standardization database is used to generate an expected pattern of response for each SUMI item. The expected pattern of response is then compared with the actual pattern obtained. Items on which there is a large discrepancy between expected and obtained patterns of response represent aspects of the system which are unique to the system being evaluated. These may be positive or negative comments on the system. Positive comments indicate the areas in which the system being evaluated may have a market advantage; negative comments indicate the areas in which more work needs to be done, or at least where there is scope for improvement. From the example of an ICA chart shown in Figure 19.2 it can be seen that there is stronger than expected support in the user profile for the statement that the system is 'clear and understandable'.

If the objective of the evaluation is to obtain a diagnosis of the software features of the product, Item Consensual Analysis may be used within a two-stage sampling

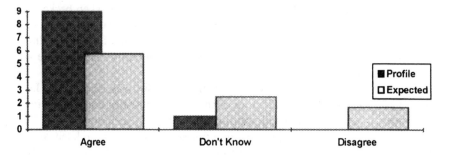

Figure 19.2 ICA chart for 'The way that system information is presented is clear and understandable'.

approach. In the first stage a large sample (i.e. more than 12 users) provides an initial SUMI analysis. The ICA outputs are used to construct an interview schedule, and in the second stage, selected users are interviewed in order to find out specific causes in the software for the extreme user reactions which emerged in the ICA analysis.

In addition to the subscales, and ICA results, each user's SUMI Global and subscale scores may be computed individually, and the use of appropriate statistics enables the evaluator to select out users for in-depth interviewing. When we use SUMI in our laboratories, we typically find that using the two-stage sampling technique yields a greater probability of actionable change suggestions than does simply interviewing the users afterwards. The SUMI items therefore act as a series of useful memory probes for the users in the evaluation.

VALIDITY OF SUMI

Three different kinds of validity study have been conducted with SUMI. First, the industrial partners within the MUSiC consortium used SUMI as part of the industry-scale validation of the MUSiC usability evaluation toolset. The objective of this exercise was to ascertain whether SUMI was usable within an industrial context and whether it yielded information that was useful to the participating companies. Both of these questions were answered positively during the course of the evaluations; a brief account is given in the MUSiC project final report (Kelly, 1994). Second, a number of laboratory-based studies have been carried out by the Human Factors Research Group; and third, studies have been carried out for industrial clients on a consultancy basis. Laboratory studies are to some extent low in ecological validity; consultancy studies are nearly always commissioned on the understanding of strict confidentiality agreements.

One important laboratory study was carried out by McSweeney (1991), who compared a number of different kinds of word processor in daily use in eleven companies within the Cork region in Ireland. The tasks studied were representative of the uses word processors would be put to in busy offices, and all participants reported having used the word processor under evaluation in their company for at least six months. Within the sample of five word processors, two products represented old technology, one was a market leader, and one a GUI-based application. SUMI did appear to discriminate between the different word processors and the SUMI subscales picked up trends that had been reported in various independent evaluations of the products (for instance, in expert evaluation reports published in popular sources such as *PC Magazine*, as well as more considered responses). Interestingly, although SUMI showed a difference in favour of the newer packages in scales such as Efficiency and Control, the two older packages still maintained high enough levels on the Helpfulness and Learnability subscales to suggest that there were reasons other than simple inertia for not replacing them.

With regard to industrial consultancies, three case study scenarios (which have been edited and idealized to protect the companies involved) are now briefly sketched to indicate the range of uses to which SUMI has been put.

Company X used SUMI to do a company-wide evaluation of all the office systems software they were using. The information from the usability profiles assisted the company in identifying software that needed to be replaced, and to draw up a staff retraining strategy. The company specifically had in mind the European Directive on Health and Safety for Work with Display Screen Equipment (EEC, 1990) when they undertook this task. SUMI enabled them to fulfil about a third of the requirements mentioned in the directive.

Company Y was about to purchase a large data entry system, and had narrowed the field down to two possibilities. The company wished to let the actual end users experience both systems to assist in the decision-making process, since this was an area in which on-site data entry was only starting to be introduced. The company, in participation with the two potential suppliers, arranged for a selection of staff to visit sites where the systems had been installed on two successive weekends. Both systems were evaluated using SUMI. The users involved felt that SUMI had sharpened their perceptions of what they should be looking for in a computer system. Although the eventual system purchased was the one with the lower usability ratings, the company used the SUMI profiles (which were exceptionally low for Learnability on the chosen system) to successfully support their bid for an enlarged training and support element in the package.

Company Z, a software vendor, had developed a new GUI version of its previously successful software package. The package was a client–server interface to a large database, and the end users of the package were using it to respond to real-time queries from the public. When the two versions of the interface were evaluated, it turned out that in many of the SUMI scales, the new version was worse than the old. After checking for familiarity effects in vain, the evaluators found that the real problem lay in the fact that in the new version the interface had become too complicated and took too long to operate in answer to a query. The new version was not released, and instead, was scheduled for a further process of redesign to take into account what had been learned about it during the evaluation.

CONCLUSIONS

It has been said that a system's end users are *the* experts in using the system to achieve goals, and that their voices should be listened to when that system is being evaluated. SUMI does precisely that: it allows quantification of the end users' experience with the software and it encourages the evaluator to focus in on issues that the end users find difficulty with (or, indeed, find exceptionally satisfying with the system). Evaluation by technical experts is also important, but it inevitably considers the system as a collection of software entities.

A questionnaire such as SUMI represents the end result of a lot of effort. The evaluator gets the result of this effort instantly when SUMI is used: the high validity and reliability rates reported for SUMI are due in large measure to the rigorous and systematic approach adopted in constructing the questionnaire and to the emphasis on industry-based testing during development. However, as with all tools, it is possible to use SUMI both well and badly. The care taken in establishing the context of use, characterizing the end user population, and understanding the tasks for which the system will be used will support sensitive testing and will yield valid and useful results in the end.

REFERENCES

DAVIES, P. and BRAILSFORD, T. (1994) *New Frontiers of Learning: Guidelines for Multimedia Courseware Developers*, Vol. 1, *Delivery, Production and Provision*, Department of Life Science, University of Nottingham, UK.

EEC (1990) Minimum safety and health requirements for work with display screen equipment (90/270/EEC), *Official Journal of the European Communities*, No. L 156, 21/6/90.

IGBARIA, M. and NACHMAN, S.A. (1991) Correlates of user satisfaction with end user computing: An exploratory study, *Information and Management*, **19**: 73–82.

IGBARIA, M. and PARASURAMAN, S. (1991) Attitudes towards microcomputers: Development and construct validation of a measure, *Int. J. Man–Machine Studies*, **35**(4): 553–573.

ISO 9241 (1992) *ISO 9241 Ergonomic Requirements for Office Work with Visual Display Terminals*, Part 10, *Dialogue Principles* (CD).

ISO 9241 (1994) *ISO 9241 Ergonomic Requirements for Office Work with Visual Display Terminals*, Part 11, *Guidance on Specifying and Measuring Usability* (Draft).

IVES, B.S., OLSON, M. and BAROUDI, J. (1983) The measurement of user information satisfaction, *Commun. ACM*, **26**: 530–545.

KELLY, M. (Ed.) (1994) *MUSiC Final Report*, Parts 1 and 2: *The MUSiC Project*, BRAMEUR Ltd, Hampshire, UK.

KIRAKOWSKI, J. (1987) The computer user satisfaction inventory, *IEE Colloquium on Evaluation Techniques for Interactive System Design*, II, London.

KIRAKOWSKI, J. and CORBETT, M. (1988) Measuring user satisfaction, in: D.M. Jones and R. Winder (Eds), *People and Computers*, Vol. IV, Cambridge: Cambridge University Press.

LEWIS, J.R. (1991) Psychometric evaluation of an after-scenario questionnaire for computer usability studies: The ASQ, *SIGCHI Bull.*, **23**(1): 78–81.

LUCEY, N.M. (1991) More than meets the I: User-satisfaction of computer systems, Diploma in Applied Psychology thesis, University College Cork, Ireland (unpublished).

McSWEENEY, R. (1991) SUMI: A psychometric approach to software evaluation, MA (Qual) thesis in Applied Psychology, University College, Cork, Ireland (unpublished).

PORTEOUS, M., KIRAKOWSKI, J. and CORBETT, M. (1993) *SUMI Handbook*, Human Factors Research Group, University College Cork, Ireland.

RAMAMURTHY, K., KING, W.R. and PREMKUMAR, G. (1992) User characteristics–DSS effectiveness linkage: An empirical assessment, *Int. J. Man–Machine Studies*, **36**: 469–505.

REDMOND-PYLE, D. and MOORE, A. (1995) *Graphical User Interface Design and Evaluation: A Practical Process*, Hemel Hempstead: Prentice Hall.

SHNEIDERMAN, B. (1992) *Designing the User Interface: Strategies for Effective Human–Computer Interaction*, 2nd edn, Reading, MA: Addison-Wesley.

WONG, G.K. and RENGGER, R. (1990) *The Validity of Questionnaires Designed to Measure User-satisfaction of Computer Systems*, National Physical Laboratory Report DITC 169/90, Teddington, Middlesex, UK.

CHAPTER TWENTY

The usability checklist approach revisited

GRAHAM I. JOHNSON

AT&T Global Information Solutions (Scotland) Ltd, Dundee, Scotland, UK

INTRODUCTION

In discussing the usability checklist approach, as presented by Ravden and Johnson (1989), it is necessary to trace the origins and the intention of the method, before addressing issues such as its relation to other, similar methods or attempting to place it in the context of the user interface design and evaluation cycle. Accordingly, this chapter first deals with its history, describes the checklist approach, and then reflects on current usage of the approach, prior to discussing some of the strengths and weaknesses of the technique.

Origins of the approach

At the MRC/ESRC Social and Applied Psychology Unit, research was undertaken on various ESPRIT (European Strategic Programme in Information Technology)-funded projects in the area of advanced manufacturing technology (see e.g. Johnson and Wilson, 1988). One of the major outcomes of studying and developing user interfaces for systems within this area was the checklist approach to usability evaluation (Johnson, 1989; Johnson *et al.*, 1989; Ravden and Johnson, 1989). The basis for this work was a combination of practical experience in the evaluation of human–computer interfaces at various stages of development and the emerging guidelines literature (e.g. Clegg *et al.*, 1988; Gardner and Christie, 1987). The process and checklist tool was developed iteratively with our industrial partners under the auspices of this research.

Overview of the Ravden and Johnson approach

This 'off-the-shelf' method is best known for the checklist that is central to the approach. As a type of method, Lindgaard (1994) considers this approach to be a global usability analysis: '... an alternative method in which people from different disciplines may fruitfully take part; it is a hybrid of heuristic evaluations and pre-designed tasks, although performance is not measured, but the evaluators integrate task experience into the usability analysis as part of the overall impression' (p.143).

In this approach the checklist is similar to a self-report questionnaire used by evaluators, and employs nine criterion-driven sections:

Visual clarity
Consistency
Compatibility
Informative feedback
Explicitness
Appropriate functionality
Flexibility and control
Error prevention and correction
User guidance and support.

Each of these sections contains detailed questions related to each specific criterion. For instance, within the Visual clarity section, questions such as 'Do screens appear uncluttered?' appear together with a standard, simple response scale ('always'; 'most of the time'; 'some of the time'; 'never'). These criterion-based sections are followed by two general sections, the first of which uses a closed-question format where checklist users respond to the categories 'major problems'/'minor problems'/ 'no problems'. A variety of topics is included in this section such as 'unexpected actions by the system', 'knowing what to do next', 'having to remember too much information', and so on. A brief section containing open-ended general questions on system usability concludes the checklist. More than 150 main items are used within the checklist.

Japanese translations of the checklist approach were published in 1993, and have apparently found regular use in industrial and academic settings.

Intended scope and usage of the approach

Our primary aim in authoring the method was to produce a straightforward and comprehensive technique that could be used in practical assessments of human–computer interfaces. In other words, the chief target audience was development engineers and system designers without backgrounds in human factors or usability engineering – people within industry who need to carry out basic usability evalua-tions as part of a development cycle. Our intention was to enable representative end users to complete the checklist in basic post-trial settings. It was the high face validity and the pragmatic nature of the 'evaluation tasks and checklist' approach that seemed to differentiate it from both volumes of guidelines (e.g. Smith and Mosier, 1986), and theoretical treatments of the area (e.g. Norman and Draper, 1986). As for the intended usage of the checklist, it was recognized from the outset that it would be adapted or used only partially, according to the stage within the development cycle, evaluator type, product, prototype, concept or system type, and so on. We had not envisaged its widespread use as a tool for ergonomists and usability professionals engaged in brief interventions and expert evaluations.

CURRENT USES OF THE USABILITY CHECKLIST APPROACH

From a recent informal survey of the usage of the Ravden and Johnson approach conducted by the author, it appears, predictably, that the checklist is being used in a wide variety of situations.

Users groups: an informal survey

In nearly all cases, teaching being an exception, the checklist has been taken up as convenient and straightforward technique. In general, those who have made practical use of the approach have been encouraged to use it again.

Clearly, reports of the use of this checklist approach do not always appear in the public domain, especially where consultants or in-house groups are involved in confidential work. In research and development, the approach has been adopted in two distinct areas of activity: in carrying out (system) usability evaluations, and in method and process development. The checklist has also been employed as courseware for those from computing and engineering development backgrounds, offered in part as a tool, but also as an illustration of part of the 'contents' of a usability investigation.

In the author's survey of the uptake and current uses of the R&J checklist approach, three sources were used: first, feedback from publishers, which could give only an idea of the distribution of sales (by region and by type) and, of course, says little about its use (or non-use); second, a brief search of the contemporary human factors literature to provide some indication of the scope of the use of the R&J checklist approach; and thirdly, and perhaps most usefully, personal enquiries by means of an informal telephone survey (mostly within the UK) of companies and groups using the R&J checklist. The findings are described briefly below.

Usability evaluation

With respect to evaluation exercises, as opposed to process augmentation, there is a variety of published examples of the use of the checklist approach in straightforward usability evaluation, including:

- user interfaces in hospital systems (Forrester, 1994),
- consumer product design (Johnson and van Vianen, 1993),
- word-processing software packages (Smythe, 1992),

- advanced telephone system interfaces (Jordan and Kerr, 1993).

Many organizations make occasional or regular use of the usability checklist approach; these include IBM UK (for various projects), ICL Fujitsu UK (in user interface evaluations), Hoogovens, NL (for process control interfaces), BNFL (for interface design generally), AT&T GIS, UK (various projects), Philips, NL (various), Customs & Excise UK (internal systems interfaces), Sports Council UK (use with database front-ends for lottery fund allocation), AIT UK (in the evaluation of public interfaces), KPMG Peat Marwick, London (various), Unilever UK (various UI projects), Telecom Australia (user interface evaluation), the UK Home Office, and several other large IT companies and financial institutions.

The approach is currently being used to evaluate a wide variety of product and system types, perhaps more than were originally envisaged as being within the scope of the checklist. Also, it is clear that the method is used at early and late stages in the development cycle. More importantly, the R&J approach is commonly being used by human factors professionals as an aid to expert and heuristic evaluation. Rather than have 'subjects' complete the checklist, usability practitioners are making use of it themselves for evaluation purposes.

Usability evaluation method development

As might be expected, the usability checklist has also found itself the subject of further development or has been used in combination with other techniques, such as

- within EVADIS II (see Reiterer, 1992; Reiterer and Oppermann, 1993),
- within general user interface evaluation approaches (e.g. Parsons and Sparshatt, 1991; Tainsh, 1994),
- in combination with existing analysis and evaluation approaches, e.g. STUDIO (Browne, 1994).

Typically, in these contexts the checklist is used in adapted form as part of the evaluation stages. Within the MRC/ESRC Social and Applied Psychology Unit, Sheffield, some further research on the use of the checklist is now being conducted with the goal of establishing the relationships between checklist items, and with certain behavioural measures.

It should also be noted that the R&J checklist has had reasonable exposure via courses and tutorials. It is included in human–computer interaction modules within undergraduate courses, and has been a component of a recent Ergonomics Society Annual Conference tutorial (Stanton and Baber, 1994).

RELATIONSHIP TO HEURISTIC EVALUATION APPROACHES

As 'discount usability engineering' becomes well publicized (see Nielsen, 1993), the uptake of this category of method – Lindgaard's (1994) 'global usability analyses', or Nielsen's 'usability inspection methods' – seems more and more likely, if one assumes that usability issues are receiving increasing attention within product and system design. The general appeal of heuristic evaluation and similar techniques is obvious in that they require evaluators, often from different disciplines, rather than numerous participants for controlled, monitored usability trials, and the concomitant cost in terms of time.

In terms of content, it is not surprising that there are large overlaps between guidelines, heuristics as used in evaluation, and the criteria and items that appear within the usability checklists. As they are generally derived from the same or similar sources, then we would naturally expect the same main criteria to appear. 'Consistency', 'Visibility', and 'Error prevention' are good examples of this overlap. One main advantage of the checklist approach with its many items is that these *a priori* categories can be readily used to 'prime' evaluators, and the meaning of the criteria labels is supported by the items themselves. In contrast, heuristic approaches offer little detail of this kind. The general level at which heuristics are described (and then used) in usability evaluation seems to be both an advantage (simplicity and time-saving) and a disadvantage (possible interpretation problems and lack of precision). As with heuristic evaluation approaches, it is highly likely that the checklist approach will capture different usability problems from those identified by approaches such as formal methods, theory-based performance models, and intensive user-based trials. It is reasonable to suppose that, as with heuristic evaluation techniques, the checklist may catch comparatively more 'less severe' problems. Similarly, in advising on the numbers of evaluators one may want to use with the checklists, we may well look at the guidance given for heuristic techniques. However, one obvious difference between the usability checklist approach and heuristic evaluation is the ability of the checklist to discriminate between areas as a result of the response scales attached to items in the checklist.

These inspection techniques are usually considered complementary to more labour-intensive techniques such as user trials within usability laboratories. It is perhaps too simplistic to assume that the early stages of user interface design can be dominated by heuristic evaluations, the checklist approach or walk-throughs. A large number of factors need to be taken into account – robustness of the prototype, intended users, purpose of the system, and context of use, to name but a few.

USABILITY CHECKLIST APPROACH ISSUES

This section discusses a number of current issues in the use and further development of the method. Many of these issues are being faced by most techniques for usability evaluation. To a large extent, the checklist approach is limited in the same general ways as are many other guidelines (see Bevan and Macleod, 1994, for a discussion of these limitations).

User feedback

First, we present some of the concerns and comments mentioned by current users during the survey. In practising what we preach, in the informal survey of users we sought not only information about the applications of the approach, but also the concerns and comments of the end users. In general, most comments were favourable, and all current users will or would use the checklist approach again. Summarizing their views on the positive aspects of the checklist, the users consider it to be highly practical; visible to clients and evaluators in terms of its purpose; relatively quick to administer; relatively comprehensive; and flexible and undemanding in analysis. The concerns raised about the checklist include its paper format; its length in terms of the time required when used fully; the wording, which is often tailored to specific or local applications; and the lack of a standard method of analysis. Clearly, many trade-offs need to be made when applying the usability checklist, and many of its strengths are also potential weaknesses; for instance, the exhaustive nature of the checklist inevitably means that it is very long.

Reflecting a definition of usability

The nature of the checklist described here, and of related tools such as SUMI (Software Usability Measurement Inventory; Kirakowski, 1994), often gives rise to discussions about operational definitions of usability, which decompose the construct of 'usability' into its associated attributes. For instance, Jordan (1994) refers to the checklist as describing those 'various properties which influence usability of an interface' (p.457). The issue pertains to the extent to which we consider the content of these tools to reflect our understanding of usability. The implication with these techniques (and also with heuristic approaches) is that if a system 'possesses' these qualities then it can be pronounced usable.

Convenience

One of the main characteristics of this type of method is its convenience. Not only is it an 'off-the-shelf' technique, it is also easy to understand and straightforward to apply. The amount of instruction to evaluators is fairly minimal and the analysis options are loosely specified, facilitating a fair degree of both convenience and freedom. Criteria put forward for the use of guidelines (e.g. Johnson et al., 1986) still hold true today in assessing the convenience of methods.

Task scenario dependency

It is sometimes overlooked that the application of the usability checklists to prototypes, systems or product interfaces is dependent on the use of the whole method, which incorporates basic task analysis and functionality familiarity, followed by definitions of specific, representative tasks. As with all methods within usability engineering that rely upon the specification of tasks to be carried out by intended end users or specified evaluators, full attention needs to be given to these aspects, and the associated evaluation of the tasks in pilot studies.

Context

As the human factors community becomes increasingly aware of the need to address context (physical, cultural, organizational, etc.) when looking at usability issues, the same holds true for those methods we use within design and development. To a large extent, checklist and usability heuristic approaches are used as 'stand-alone' tools. On the whole, they are not particularly sensitive to context, although they could, in principle, be augmented to better accommodate contextual factors.

Developer education and specification

A by-product of the use of the checklist with developers who have little or no human factors background is that the checklist can assist in promoting issues critical to the usability of a user interface. In a similar vein, the use of the checklist early in the development cycle, together with target values, can be of great value in setting usability goals or writing specifications for products.

Subjective versus objective aspects of usability

One obvious area of contention for any subjective technique is its relationship with behavioural measures. Although it may be common for usability practitioners to conduct user trials with measured behaviours (say, errors encountered) and to follow up with a post-trial questionnaire, the relationship between the results is often unclear. We cannot assume that people are especially good judges of their own behaviour, and it is always possible that there will be discrepancies between subjective and objective measures.

Scope

There will often be debate as to the scope of techniques with respect to a product or user interface type, and with respect to the phase within the design cycle. We can discuss the application of the R&J checklist in both cases.

Intuitively, one would expect to be able make use of checklists and the like only when a preliminary product design is available, and thereafter. However, it is quite possible to make use of checklist tools early within the product design cycle (as mentioned earlier) as an aid to specification. Similarly, products within which a screen has only limited importance would test the scope of such a usability checklist tool. For obvious reasons the checklist has seen most applications with desktop software products and bespoke systems that rely upon VDUs. When one considers the types of product that contain, say, two-line liquid crystal displays (whether these serve consumer or professional products) the applicability of the checklist is reduced. However, the guiding principles (i.e. the criteria from which core sections of the checklist are drawn) remain the same. As we move further away from screen-based products, say, to complex telephone systems, the checklist is also more cumbersome to use and requires customization.

Within the emerging world of multimedia and 'edutainment', although the applications are screen-based, it is difficult to imagine a static checklist that could hope to easily cover the range of user experiences that may make up interaction with such products. This challenge, however, is common to the majority of, if not all, general human factors techniques that aim to assess more than the most basic performance aspects of an interaction.

ON THE FUTURE OF USABILITY CHECKLIST APPROACHES

One of the most interesting developments within the area of evaluation methodologies is the comparative studies that have been carried out. Examining the number of usability problems captured, their severity, and the associated evaluation cost is likely to yield much of use to the human factors practitioner. Studies looking at checklist approaches such as the usability checklist described above would help clarify its position relative to other techniques.

The scope of the usability checklist and related methods will always be tested by the pace of technological development and the rate at which innovative solutions enter the marketplace. For instance, dealing with the complexities of an early prototype of a video-conferencing system that incorporates advanced gesture and voice-based interaction would prove to be a challenge for many 'off-the-shelf' techniques. As noted earlier, our experience with usability checklists has pointed out not only the need for the involvement of end users within evaluation exercises, but also the ergonomists' need for tools of this type. At present it appears that their requirements for evaluation instruments are well met by checklists and the like. We can reasonably assume that the uptake of and further need for checklist-type tools can be attributed to the fact that many offer high face validity, are relatively quick and easy to apply, and are readily accessible.

Given that increasing attention is being given to general user interface usability, it is probable that the types of approach discussed here will continue to provide developers with a convenient (and inexpensive) means by which interfaces can be evaluated.

ACKNOWLEDGEMENTS

The author gratefully acknowledges AT&T Global Information Solutions (Scotland) Ltd for its support and encouragement. Thanks also to John Wilson and Jeremy Robson at University of Nottingham and David Jones at KPMG Peat Marwick, London, for their varied thoughts on this topic. The views expressed in this paper are those of the author and do not necessarily reflect those of AT&T Global Information Solutions (Scotland) Ltd.

REFERENCES

BEVAN, N. and MACLEOD, M. (1994) Usability measurement in context, *Behaviour and Information Technology*, **13**(1&2): 132–145.

BROWNE, D.P. (1994) *STUDIO: Structured User Interface Design Optimisation*, Hemel Hempstead: Prentice Hall.

CLEGG, C.W., WARR, P.B., GREEN, T.R.G., MONK, A., KEMP, N., ALLISON, G. and LANSDALE, M. (1988) *How to Evaluate Your Company's New Technology*, Chichester: Ellis Horwood.

FORRESTER, C. (1994) Case study: The usability of the human–computer interface at a regional hospital, in: *IEA '94, Proc. 12th Triennial Congress of the International Ergonomics Association*, Vol. 4, Toronto, Canada, August 1994, pp. 372–375.

GARDNER, M.M. and CHRISTIE, B. (Eds) (1987) *Applying Cognitive Psychology to User Interface Design*, London: Wiley.

JOHNSON, G.I. (1989) The user's side of the computer interface, *Applied Ergonomics*, **20**(3): 158–159.

JOHNSON, G. I. and VAN VIANEN, E.P.G. (1993) Practical experiences with consumer products, users and prototyping, in: E.J. Lovesey (Ed.), *Contemporary Ergonomics 1993*, pp. 429–434, London: Taylor & Francis.

JOHNSON, G.I. and WILSON, J.R. (Eds) (1988) *Ergonomics Matters In Advanced Manufacturing Technology*, London: Butterworth.

JOHNSON, G.I., CLEGG, C.W. and RAVDEN, S.J. (1986) Human factors in a flexible assembly system: Design guidelines for design guidelines, in: D.J. Oborne (Ed.), *Contemporary Ergonomics 1986*, pp. 78–82, London: Taylor & Francis.

JOHNSON, G.I., CLEGG, C.W. and RAVDEN, S.J. (1989) Towards a practical method of user interface evaluation, *Applied Ergonomics*, **20**(4): 255–260.

JORDAN, P.W. (1994) What is usability?, in: S.A. Robertson (Ed.), *Contemporary Ergonomics 1994*, pp. 454–458, London: Taylor & Francis.

JORDAN, P.W. and KERR, K. (1993) A multifunction 'phone system evaluation, in: E.J. Lovesey (Ed.), *Contemporary Ergonomics 1993*, pp. 416–421, London: Taylor & Francis.

KIRAKOWSKI, J. (1994) *SUMI: Software Usability Measurement Inventory*, Paper presented at 'Usability Evaluation in Industry', September 1994, Eindhoven.

LINDGAARD, G. (1994) *Usability Testing and System Evaluation: A Guide for Designing Useful Computer Systems*, London: Chapman & Hall.

NIELSEN, J. (1993) *Usability Engineering*, New York: Academic Press.

NIELSEN, J. (1994) Enhancing the exploratory power of usability heuristics, in: B. Adelsen

S. Dumais and S. Olson (Eds), *Proc. Conf. CHI'94: Human Factors in Computing Systems*, ACM, pp. 152–158.

NORMAN, D.A. and DRAPER, S.W. (Eds) (1986) *User-centred System Design: New Directions in Human–Computer Interaction*, Hillsdale, NJ: Lawrence Earlbaum Associates.

PARSONS, K. and SPARSHATT, J. (1991) A toolkit for the user evaluation of software interfaces, in: E.J. Lovesey (Ed.), *Contemporary Ergonomics 1991*, pp. 73–78, London: Taylor & Francis.

RAVDEN, S.J. and JOHNSON, G.I. (1989) *Evaluating Usability of Human–Computer Interfaces: A Practical Method*, Chichester: Ellis Horwood.

REITERER, H. (1992) EVADIS II: A new method to evaluate user interfaces, in: A. Monk, D. Diaper and M.D. Harrison (Eds), *People and Computers VII, Proc. HCI'92 Conference*, Cambridge: Cambridge University Press.

REITERER, H. and OPPERMANN, R. (1993) Evaluation of user interfaces: EVADIS II: A comprehensive evaluation approach, *Behaviour and Information Technology*, **12**(3): 137–148.

SMITH, S.L. and MOSIER, J.N. (1986) *Guidelines for Designing User Interface Software*, Report No. MTR-10090 ESD-TR-86-278, MITRE Corporation, Bedford, MA, USA.

SMYTHE, J.A. (1992) Evaluating software packages: A case study, in: E.J. Lovesey (Ed.), *Contemporary Ergonomics 1992*, pp. 385–390, London: Taylor & Francis.

STANTON, N. and BABER, C. (1994) *A Pragmatic Approach to the Design and Evaluation of User Interfaces*. Tutorial presented at the Ergonomics Society Annual Conference, 'Ergonomics for All', University of Warwick, UK, 19–22 April 1994.

TAINSH, M.A. (1994) Human factors contributions to the acceptance of computer supported systems, in: S.A. Robertson (Ed.), *Contemporary Ergonomics 1994*, pp. 155–160, London: Taylor & Francis.

CHAPTER TWENTY ONE

SUS: a 'quick and dirty' usability scale

JOHN BROOKE

Redhatch Consulting Ltd, Earley, Reading, UK

USABILITY AND CONTEXT

Usability is not a quality that exists in any real or absolute sense. Perhaps it can be best summed up as being a general quality of the *appropriateness to a purpose* of any particular artefact. This notion is neatly summed up by Terry Pratchett in his novel *Moving Pictures*:

> 'Well, at least he keeps himself fit,' said the Archchancellor nastily. 'Not like the rest of you fellows. I went into the Uncommon Room this morning and it was full of chaps snoring!'
>
> 'That would be the senior masters, Master,' said the Bursar. 'I would say they are supremely fit, myself.'
>
> '*Fit*? The Dean looks like a man who's swallered a bed!'
>
> 'Ah, but Master,' said the Bursar, smiling indulgently, 'the word "fit", as I understand it, means "appropriate to a purpose", and I would say that the body of the Dean is supremely appropriate to the purpose of sitting around all day and eating big heavy meals.' The Dean permitted himself a little smile. (Pratchett, 1990)

In just the same way, the usability of any tool or system has to be viewed in terms of the context in which it is used, and its appropriateness to that context. With particular reference to information systems, this view of usability is reflected in the current draft international standard ISO 9241-11 and in the European Community ESPRIT project MUSiC (Measuring Usability of Systems in Context) (e.g. Bevan *et al.*, 1991). In general, it is impossible to specify the usability of a system (i.e. its fitness for purpose) without first defining who are the intended users of the system, the tasks those users will perform with it, and the characteristics of the physical, organizational and social environment in which it will be used.

Since usability is itself a moveable feast, it follows that measures of usability must themselves be dependent on the way in which usability is defined. It is possible to talk of some general classes of usability measure; ISO 9241-11 suggests that measures of usability should cover

- effectiveness (the ability of users to complete tasks using the system, and the quality of the output of those tasks),
- efficiency (the level of resource consumed in performing tasks), and
- satisfaction (users' subjective reactions to using the system).

However, the precise measures to be used within each of these classes of metric can vary widely. For example, measures of effectiveness are very obviously determined by the types of task that are carried out with the system; a measure of effectiveness of a word processing system might be the number of letters written, and whether the letters produced are free of spelling mistakes. If the system supports the task of controlling an industrial process producing chemicals, on the other hand, the measures of task completion and quality are obviously going to reflect that process.

A consequence of the context-specificity of usability and measures of usability is that it is very difficult to make comparisons of usability across different systems. Comparing usability of different systems intended for different purposes is a clear case of 'comparing apples and oranges' and should be avoided wherever possible. It is also difficult and potentially misleading to generalize design features and experience across systems; for example, just because a particular design feature has proved to be very useful in making one system usable does not necessarily mean that it will do so for another system with a different group of users doing different tasks in other environments.

If there is an area in which it is possible to make more generalized assessments of usability, which could bear cross-system comparison, it is the area of subjective assessments of usability. Subjective measures of usability are usually obtained through the use of questionnaires and attitude scales, and examples exist of general attitude scales which are not specific to any particular system (for example, CUSI; Kirakowski and Corbett, 1988).

INDUSTRIAL USABILITY EVALUATION

The demands of evaluating usability of systems within an industrial context mean that often it is neither cost-effective nor practical to perform a full-blown context analysis and selection of suitable metrics. Often, all that is needed is a general indication of the overall level of usability of a system compared to its competitors or its predecessors. Equally, when selecting metrics, it is often desirable to have measures which do not require vast effort and expense to collect and analyze data.

These sorts of considerations were very important when, while setting up a usability engineering programme for integrated office systems engineering with Digital Equipment Co. Ltd, a need was identified for a subjective usability measure. The measure had to be capable of being administered quickly and simply, but also had to be reliable enough to be used to make comparisons of user performance changes from version to version of a software product.

The need for simplicity and speed came from the evaluation methods being used; users from customer sites would either visit a human factors laboratory, or a travelling laboratory would be set up at the customer site. The users would then work through evaluation exercises lasting between 20 minutes and an hour, at the end of which a subjective measure of system usability would be collected. As can be imagined, after this period of time, users could be very frustrated, especially if they had encountered problems, since no assistance was given. If they were then presented with a long questionnaire, containing in excess of 25 questions, it was very likely that they would not complete it and there would be insufficient data to assess subjective reactions to system usability.

SUS: THE SYSTEM USABILITY SCALE

In response to these requirements, a simple usability scale was developed. The System Usability Scale (SUS) is a simple, ten-item scale giving a global view of subjective assessments of usability.

SUS is a *Likert scale*. It is often assumed that a Likert scale is simply one based on forced-choice questions, where a statement is made and the respondent then indicates the degree of agreement or disagreement with the statement on a 5 (or 7) point scale. However, the construction of a Likert scale is somewhat more subtle than this. Whilst Likert scales are presented in this form, the statements with which the respondent indicates agreement and disagreement have to be selected carefully.

The technique used for selecting items for a Likert scale is to identify examples of things which lead to extreme expressions of the attitude being captured. For instance, if one was interested in attitudes to crimes and misdemeanours, one might use serial murder and parking offences as examples of the extreme ends of the spectrum. When these examples have been selected, then a sample of respondents is asked to give ratings to these examples across a wide pool of potential questionnaire items. For instance, respondents might be asked to respond to statements such as 'hanging is too good for them', or 'I can imagine myself doing something like this'.

Given a large pool of such statements, there will generally be some where there is a lot of agreement between respondents. In addition, some of these will be ones where the statements provoke extreme statements of agreement or disagreement among all respondents. It is these latter statements which one tries to identify for inclusion in a Likert scale, since, we would hope that, if we have selected suitable examples, there would be general agreement of extreme attitudes to them. Items where there is ambiguity are not good discriminators of attitudes. For instance, while one hopes that there would be a general, extreme disagreement that 'hanging is too good' for those who perpetrate parking offences, there may well be less agreement about applying this statement to serial killers, since opinions differ widely about the ethics and efficacy of capital punishment.

SUS was constructed using this technique. A pool of 50 potential questionnaire items was assembled. Two examples of software systems were then selected (one a linguistic tool aimed at end users, the other a tool for systems programmers) on the basis of general agreement that one was 'really easy to use' and one was almost impossible to use, even for highly technically skilled users. Twenty people from the office systems engineering group, with occupations ranging from secretary through to systems programmer then rated both systems against all 50 potential questionnaire items on a five-point scale ranging from 'strongly agree' to 'strongly disagree'.

The items leading to the most extreme responses from the original pool were then selected. There were very close intercorrelations between all of the selected items (±0.7 to ±0.9). In addition, items were selected so that the common response to half of them was strong agreement, and to the other half, strong disagreement.

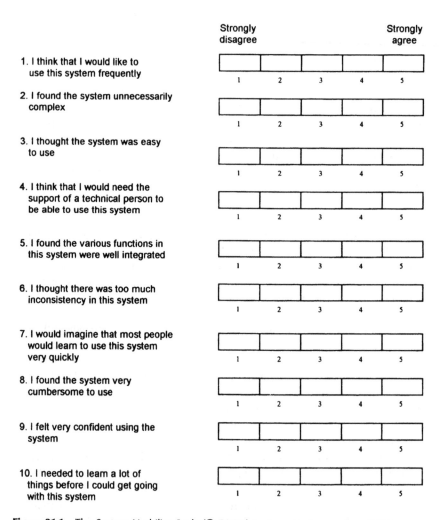

Figure 21.1 The System Usability Scale (© Digital Equipment Corporation, 1986).

This was done in order to prevent response biases caused by respondents not having to think about each statement; by alternating positive and negative items, the respondent has to read each statement and make an effort to think whether they agree or disagree with it.

The System Usability Scale is shown in Figure 21.1. It can be seen that the selected statements actually cover a variety of aspects of system usability, such as the need for support, training, and complexity, and thus have a high level of face validity for measuring the usability of a system.

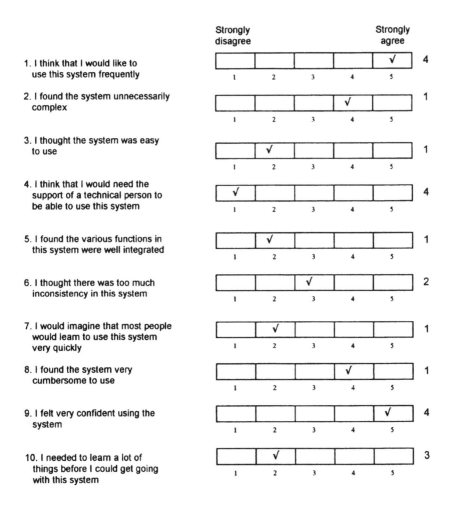

Total score = 22

SUS Score = 22 *22.5 = 55

Figure 21.2   The System Usability Scale (© Digital Equipment Corporation, 1986).

Using SUS

The SU scale is generally used after the respondent has had an opportunity to use the system being evaluated, but before any debriefing or discussion takes place. Respondents should be asked to record their immediate response to each item, rather than thinking about items for a long time. All items should be checked. If a respondent feels that they cannot respond to a particular item, they should mark the centre point of the scale.

Scoring SUS

SUS yields a single number representing a composite measure of the overall usability of the system being studied. Note that scores for individual items are not meaningful on their own.

To calculate the SUS score, first sum the score contributions from each item. Each item's score contribution will range from 0 to 4. For items 1,3,5,7 and 9 the score contribution is the scale position minus 1. For items 2,4,6,8 and 10, the contribution is 5 minus the scale position. Multiply the sum of the scores by 2.5 to obtain the overall value of SU. SUS scores have a range of 0 to 100. Figure 21.2 shows an example of a scored SU scale.

CONCLUSION

SUS has proved to be a valuable evaluation tool, being robust and reliable. It correlates well with other subjective measures of usability, such as the general usability subscale of the SUMI inventory developed in the MUSiC project (Kirakowski, personal communication). SUS has been made freely available for use in usability assessment, and has been used for a variety of research projects and industrial evaluations; the only prerequisite for its use is that any published report should acknowledge the source of the measure.

ACKNOWLEDGEMENT

SUS was developed as part of the usability engineering programme in integrated office systems development at Digital Equipment Co. Ltd, Reading, UK.

REFERENCES

BEVAN, N., KIRAKOWSKI, J. and MAISSEL, J. (1991) What is usability?, in: H.-J. Bullinger (Ed.), *Human Aspects in Computing: Design and Use of Interactive Systems and Work with Terminals*, Amsterdam: Elsevier.
KIRAKOWSKI, J. and CORBETT, M. (1988) Measuring user-satisfaction, in: D.M. Jones and R. Winder (Eds), *People and Computers IV*, Cambridge: Cambridge University Press.
PRATCHETT, T. (1990) *Moving Pictures*, London: Gollancz.

CHAPTER TWENTY TWO

FACE: a rapid method for evaluation of user interfaces

REINOUD HULZEBOSCH

Informaat, 1200 AA Hilversum, The Netherlands

ANTHONY JAMESON

Universität des Saarlands, D-66041 Saarbrücken, Germany

INTRODUCTION

There are many situations in which it is desirable to perform an empirical evaluation of a user interface. Doing this rapidly yet reliably is a difficult challenge. In this chapter we present the design of a method for the fast evaluation, analysis, and reporting of the usability of a computer interface from a cognitive ergonomics' viewpoint.

The purpose of the method is to enable the examination of the usability of a large number of different computer interfaces in work environments. The result of a usability evaluation lists the shortcomings of the user interface and contains general conclusions about usability. Prior to the design of the method, four objectives were determined:

- Fast applicability: applying the method should not produce high costs, implying that cognitive ergonomists should be able to complete an evaluation as quickly as possible. One evaluation should take no more than four hours of evaluation and four hours of analysis and writing a report.

- Valid outcome: the resulting report must contain observations and conclusions that are of good quality from the cognitive ergonomics' viewpoint: evaluation techniques should be applied correctly, constituting a complete as possible set of real usability problems.

- Wide applicability: the applicability of the method should not be restricted to certain interfaces. Evaluation criteria should therefore represent usability of most computer interfaces and should be easy to update.

- Thoroughness: the evaluation must be thorough enough to discover all problem areas of the target system. The focus of the evaluation should also cover organizational aspects of computer use, in addition to cognitive aspects.

BRIEF OVERVIEW OF THE METHOD

The evaluation method FACE (Fast Audit based on Cognitive Ergonomics) supports experts in the field of cognitive ergonomics at several points in applying an evaluation, for example, in getting to know the evaluation situation, in deciding which evaluation techniques to use, and in applying the evaluation (see Hulzebosch, 1992). Furthermore, it provides standard forms, for example checklists and a report, and creates a database of evaluation data to enable updating the method. FACE is implemented in HyperCard 2.0. The procedure for applying FACE is as follows.

The first phase of the evaluation method is the preparation phase. Through the use of three short checklists filled in by the client (organizational user and end user), evaluators as well as users are prepared for the evaluation: the evaluators can estimate the situation of computer use and the context of evaluation, and users are introduced to aspects and questions they can expect during the evaluation.

The second phase of the evaluation method is the evaluation phase, where evaluators design and apply the evaluation. Based on the specific situation, FACE suggests an evaluation design, proposing evaluation techniques that are fast, valid and applicable in the specific situation, and pointing out aspects that need special attention, ensuring that the evaluation is in correspondence with the objectives of the method.

Having chosen a set of techniques, the evaluator can print documents that are supportive during the evaluation, for example, lists containing points of attention, usability dimensions (e.g. consistency, language use or compatibility), and time schedules. FACE puts all data and selections into a database, and informs the evaluator how to adjust the chosen techniques to the specific situation. With the printed documents and the adjusted material, the evaluator carries out the evaluation.

The third and last phase is the presentation phase, which includes the analysis of the data as well as the writing of a report. FACE proposes a strategy for analyzing the data gathered. Through using a word processor in the outline mode, evaluators can quickly locate, enter and (re)structure the data to gain an overview of all data and to interpret them to draw conclusions. Finally, FACE pastes evaluation data in a standard report, and presents it as a word processor file. The evaluator can paste in the methodological aspects, observations, and conclusions, finishing the evaluation.

In the long run, the checklists, filed reports, and the evaluation database will help in gaining insight into the most frequently occurring usability and field research problems. Based on this insight, the method can be updated to meet its objectives.

DESIGN RECOMMENDATIONS FOR FACE

A review of the existing literature yielded three recommendations concerning the design of FACE: efficiency, multiple perspectives and tailoring:

Efficiency:
- The evaluation method should provide templates to guide the evaluation and reduce the amount of information.
- Usable evaluation techniques must be fast and inexpensive, systematic, and efficient.
- Background information should be directly accessible to the evaluator.
- Measures and techniques should be relevant to the questions of the user of the evaluation information and appropriate for the evaluation criteria.

Multiple perspectives:
- The evaluation should have a modular approach to allow examination of a user interface from several points of view.
- The evaluation method should give the evaluator information about the properties of a wide range of evaluation techniques and help to determine their applicability to the situation at hand.
- The method should emphasize the selection of evaluation techniques that elicit judgements from both evaluator and end user.

Tailoring:
- The evaluation should be aimed at the interest of the specific persons who will be making use of the evaluation results (e.g. end users, managers).
- The evaluation method should facilitate the construction of customized lists of usability dimensions, questions to be asked, and guidelines to be considered.
- The purpose of the evaluation should be discussed in advance and background information should be collected on both system and organization.
- The evaluation should focus on real-use situations within the organizational context.
- The evaluation should be well planned to increase the chances that the information will be utilized.

OVERVIEW OF EXPERIENCES WITH EVALUATION TECHNIQUES

The literature survey yielded a compilation of information about various specific evaluation techniques, including system walk-through, heuristic evaluation, questionnaire and checklist, thinking aloud, interview, group discussion, naturalistic observation, picture presentation, audio recording, and video recording. This yielded, for each technique, the following types of information: (a) points to attend

Table 22.1　Considerations relevant to the selection of evaluation techniques

Evaluation technique	Considerations
System walk-through	■ The system must be available during the evaluation. ■ The system must not be too large or complex. ■ Apply if no other technique with the viewpoint of an expert is included. ■ Inclusion of an additional user test is necessary. ■ If desired, suitable for testing hypotheses. ■ Usability dimensions must be appropriate.
Heuristic evaluation	■ Three to five evaluators should be available during the evaluation. ■ Inclusion of an additional user test is necessary. ■ Is suitable when no preparation is possible. ■ Either a user or sufficient time should be available.
Questionnaire and checklist	■ Users must be available during the evaluation. ■ Requires only little introduction. ■ Set of questions must be relevant. ■ Evaluators should 'know' the users.
Thinking aloud	■ The system must be available during the evaluation. ■ Users must be sufficiently available during the evaluation. ■ Inclusion of an additional user test is necessary. ■ It should be supported with audio recording.

to in order to apply the technique rapidly yet validly; (b) an appropriate form listing documents to be consulted and filled in while applying the technique; and (c) strategies for analyzing the data generated. Table 22.1 lists some considerations that are relevant to the decision as to whether to apply a certain technique or not. Four of the ten evaluation techniques are included in this table.

The following list previews the implications of these techniques for the content and architecture of FACE:

- Emphasize discussing the purpose of the audit with the organization.
- Emphasize determining user and system availability.
- Emphasize the inclusion of both a user and an expert view in the evaluation techniques.
- Emphasize applying supportive techniques when recommended.
- Have user and system information collected at the outset.
- Recommend the inclusion of at least one test involving users.
- Support construction of a tailored list of usability dimensions.
- Support construction of a set of applicable guidelines.
- Support construction of a set of suitable questions.

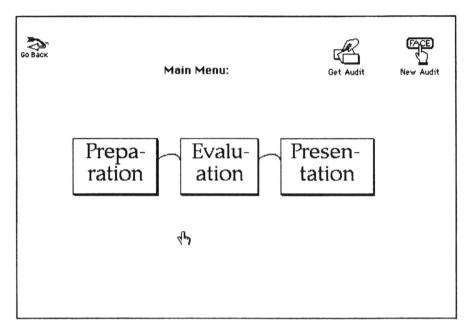

Figure 22.1   Main menu window of the HyperCard stack of FACE.

DESCRIPTION OF THE METHOD

This section presents and describes several aspects of FACE as a result of the design considerations presented above. FACE is implemented as a hypertext stack which is worked through in three phases: preparation, evaluation and presentation (see Figure 22.1).

Preparation phase

FACE supports evaluators in obtaining information about the situation of use, which will help them in designing the evaluation. Three short checklists can be used to gather information about the users, the system and the organization. The evaluator discusses the purpose of the evaluation and the possibilities for research (i.e. to what extent are system and users available) within the organization, and requests the organization to provide system documentation, i.e. a user guide and a technical manual.

The evaluator can use the system information to gain an impression of the technical characteristics of the system and the properties of the interface. Table 22.2 shows the contents and purposes of the three checklists. During the preparation phase these checklists ensure the compatibility of the evaluation with the characteristics of system use. The evaluator can construct a list of usability

Table 22.2   Contents and purposes of the three checklists in the preparation phase

Checklist	Content	Purpose
1. *User perspective* (filled in by end users of the target system)	▪ Statements relevant to usability dimensions to which users give priority values on a scale. ▪ Characteristics of the responding user.	▪ To assess the concept of usability as experienced by users. ▪ To obtain task and system characteristics. ▪ To assess the representativeness of the responding user.
2. *Organizational perspective* (filled in by a manager representing the client organization)	▪ Characteristics of the organization. ▪ Characteristics of the user population. ▪ Demands that the organization places on system use.	▪ To obtain organizational context information. ▪ To obtain information about the user population. ▪ To assess the concept of usability as experienced by the organization.
3. *Preparatory checklist* (filled in by end users of the target system)	▪ Example items from a master questionnaire considering all standard usability dimensions, to which users can answer in their own words.	▪ To introduce users to the usability questions posed during research. ▪ To encourage users to think about usability items before the research. ▪ To become familiar with the language used by users.

dimensions that fits the characteristics of the specific use situation, so as to increase the comprehensiveness of the evaluation. FACE contains a default list of 11 usability criteria: colour use, concept and metaphor, dialogue structure, documentation, error prevention, feedback, help, language use, navigation, screen design, and task performance. These dimensions of usability also appear in the master questionnaire recommended by FACE. From the information obtained in this phase evaluators can judge whether dimensions need to be added (e.g. functionality or system performance when evaluating a DTP application), or deleted (e.g. the use of colour when evaluating a system with black and white monitors), as well as which dimensions probably need special (or little) attention.

Both user and organizational perspective checklists (cf. checklists 1 and 2 in Table 22.2) increase the potential usefulness of the resulting report because the concept of usability is defined by both (a) the evaluator and (b) the organization and users. A discussion of the purpose of the evaluation and consideration of its technical limitations are also included to increase the usefulness of the evaluation.

The inclusion of the checklist focusing on the organizational perspective means that those responsible for the system design will have greater commitment to the evaluation.

The preparatory checklist (checklist 3 in Table 22.2) is a short version of the master questionnaire recommended by FACE for evaluating usability. It prepares both users and evaluator for the evaluation. First, the preparatory checklist enables the evaluator to adapt dimensions, guidelines, and questions to the specific use situation. Secondly, users are introduced to the usability questions and usability dimensions, preparing them for aspects that will be looked at, and the questions evaluators might pose during the research. Users are likely to consider the items in the checklist during their work and have answers ready before the evaluation starts.

Evaluation phase

The evaluation phase consists of a preparatory part in which the evaluator designs the evaluation by walking through the stack. FACE actively supports the evaluator in choosing a set of evaluation techniques. It prints out forms that can be filled in by hand during the evaluation and gives hints concerning the effective application of the set of techniques selected. FACE gives the information the evaluator needs to make a choice of which evaluation techniques to apply, and focuses his or her attention on aspects that need to be considered. This enables systematic consideration of usability and selection of methods and techniques, planning the evaluation and providing the evaluator with a list of steps to be followed.

FACE provides the evaluator with (i) general information about the ten evaluation techniques included, (ii) points of attention relevant to applying the techniques rapidly, leading to a valid outcome, (iii) tips derived from the study of evaluation techniques in the literature, (iv) accompanying documents, and (v) strategies on how to analyze the gathered data. Together with the three checklists and the system documentation, the information base created enables the evaluator to design and plan the evaluation.

FACE distinguishes different evaluations to help the evaluator choose the most appropriate techniques. Four situational factors (roughly) determine the applicability of techniques and the possibility of fast and valid evaluation:

1. The number of evaluators available to examine the system.
2. The size and complexity of the system.
3. The extent to which users are available for testing.
4. The extent to which the normal functions of the system can be interrupted.

Having indicated these four situational factors, the evaluator is presented with suggestions for research, together with points of attention relevant to the specific situation (see Figure 22.2). All points of attention and suggestions for applying techniques are based on the considerations mentioned above. For example, if the

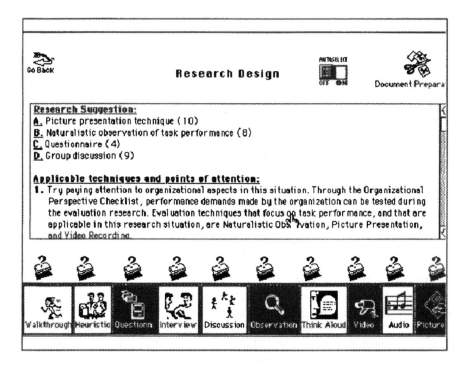

Figure 22.2   Research suggestion window of the HyperCard stack of FACE.

evaluator has rated the system not to be complex or large (situation factor 2, judged from system information obtained in the preparatory phase), FACE would recommend consideration of the picture presentation technique. However, if the evaluator has rated the availability of users (situation factor 3) as 'none', FACE would not propose this technique.

Another example is the situation in which the evaluator has rated the target system as large and complex; in this case FACE would recommend using a timetable with the list of usability dimensions. Using a timetable will prevent the evaluation from focusing too much on details and not being able to construct an integrated view of the system. Using a set of usability dimensions supports a systematic consideration of usability, taking a modular approach in evaluation and a multivariate view on usability. It reduces the influence of the personal focus of the evaluator.

A third example is the recommendation to select a set of techniques that (i) takes in both user and evaluator perspective, (ii) includes supportive techniques, and (iii) does not take too much time (approximately three hours). To remind the evaluator of this, FACE highlights supportive techniques when certain techniques have been chosen by the evaluator, and adds up the estimated time for applying the selected techniques in the specific research situation, warning the evaluator when the estimated time for application of the selected set of techniques exceeds three hours.

Nevertheless, the evaluator has a good deal of freedom in designing the research, since otherwise it might be carried out with inadequate motivation, possibly leading to a useless report. Eclectic evaluation is considered preferable to evaluations that are constrained by preconceived notions.

For each technique finally selected, FACE prompts the evaluator to print the accompanying documents. For instance, if the evaluator has chosen to apply a system walk-through, he or she is prompted to adapt the standard list of guidelines, questions, and usability dimensions to the specific situation of research. A standard set of usability dimensions, including those used in the preparatory checklists, is available by using a word processor. It is easily modifiable in the outlining mode. The usability dimensions are compatible with an electronic usability questionnaire used by the company that will introduce FACE.

When the evaluator has completed the research design, FACE enters the research design and specific situation variables into the database, thus ensuring that long-term adjustments to FACE will be based on the most frequently occurring research situations. The word processor file containing the selected set of usability dimensions is printed and saved. The printed file will be used by the evaluator during the evaluation to note down observations.

The HyperCard stack of FACE gives no support during the actual research, apart from presenting the accompanying documents. The evaluator carries out the evaluation, keeping these documents within reach, and noting down observations and findings near the relevant usability dimension of the printed file mentioned above. Timetables or, for example, user characteristics can be inspected directly, to keep in mind user characteristics during the research.

Presentation phase

The last phase of FACE is the presentation phase, which the evaluator reaches after having completed the evaluation research. Analysis of the data and construction of the report are the two steps in this phase. FACE opens the word processor file that was saved before the evaluation research.

The evaluator enters the data that were noted on the printed version of this file. FACE recommends subsequently dragging the entered data to relevant usability dimensions in outlining mode, creating an overview of all the data gathered by different techniques. All data, including for example evaluator characteristics in case of involvement of more than one evaluator, can be analyzed and used to draw conclusions. At this stage, FACE gives no support, because the interpretation as well as drawing conclusions from the data requires careful consideration of various aspects of the research, which only the evaluator is expected to be able to do (considering the state of the art).

The final step of the presentation phase is the creation of an audit report. FACE contains a standard report into which research procedures, research data, and conclusions can be pasted. The HyperCard stack automatically fills in audit data, such as names and dates, into the standard report. The evaluator is encouraged to give conclusions at the level of usability dimensions. The content of the resulting report should, however, depend on the purpose of the evaluation agreed upon in the preparation phase, and the technical limitations of the system, to provide useful information.

IMPROVEMENTS TO THE METHOD

Experience from field evaluations with FACE has shown that evaluators show few differences in their selections of evaluation techniques. The suggestions given by FACE allow evaluators room to select the techniques they prefer, within a certain range of applicability. The time an evaluator spends is about eight hours. The tips given by FACE are considered useful, as is the computer support. FACE needs further improvement of the standard usability dimensions included: some dimensions overlap, and others are missing. Further improvements are needed at the stage of report creation; in particular, pasting result and conclusions would speed up the evaluation. Finally, the time schedules seem to leave too much room for evaluators to give too much attention to certain dimensions while neglecting others.

REFERENCE

HULZEBOSCH, R. (1992) Not just a pretty FACE: A computer-supported method for fast auditing of user interfaces, MSc thesis, University of Nijmegen, Department of Cognitive Science (available from the author).

PART SEVEN

Task Analysis

CHAPTER TWENTY THREE

Hierarchical task analysis: an overview

R.B. STAMMERS

Psychology Group, Aston University, Birmingham, UK

INTRODUCTION

Hierarchical task analysis (HTA) was first described in the literature by Annett and Duncan (1967), although the general approach had been devised earlier by John Annett. The context in which it was initially used was a training one, but it has become an approach that has been applied in a number of human factors roles such as interface design and human reliability. It has also been used in a number of industries. HTA is probably best known for its use in industrial process control (e.g. Piso, 1981; Astley and Stammers, 1987; Shepherd, 1993), but it has been used for a great variety of other tasks including air traffic control, sports skills and for human–computer interaction.

HTA is best seen as a general approach to analyzing tasks rather than as a tightly formulated methodology. As such, the key features are a hierarchical representation of a task, flexibility in the level of detail of information collected, and a tailoring of the analysis to the purpose in hand. There has never been any strict rules on the details of the approach other than the observance of some guiding principles. One problem that the approach has had is the misinterpretation of some of its features. The most serious example of this is when it is assumed that the hierarchy represents a model of cognitive activity. This is not the intention. Rather, it is felt that the hierarchy is a convenient form of representation of a complex entity, a task. The issue of the hierarchy being a convenience rather than a theoretical point draws attention to the approach being a task-based one. The focus is on the goals of the user rather than on the details of the technology being used. It also means that the analyst does not have to refer to any particular theory of human performance.

Another problem has been that descriptions of the approach have not been widely available. The later criticisms do not now apply, since the approach has been described in Kirwan and Ainsworth (1992) and specifically discussed in the HCI context by Shepherd (1989).

It is probably best to begin by an examination of the basic principles of HTA. The first is that a task can be conveniently represented as a hierarchical description. This description has at its top an overall statement of the task in a verb–noun form, e.g. 'change a car wheel'. This overall description is then broken down, into more or less detail, by a process of progressive redescription. The units of analysis are termed 'operations' and are used for subunits at all levels of the hierarchy. All operations take the form of an instruction to carry out an activity. This feature avoids the problem of having to choose terms for the units at different levels of analysis, e.g. 'elements, actions, units', and allows for varying levels of description both within a single analysis and across analyses.

A task of any complexity will be broken down into a number of operations after the initial description. Thus for changing a car wheel there might be, 'position car', 'locate tools and spare wheel', 'remove wheel', and 'replace wheel'. Each of these operations needs to be examined in turn, and, if necessary, broken down further into sub-operations. Thus 'position car' will be concerned with locating a safe and convenient site to change the wheel, and might refer to whether the hand-brake is to be applied, or which gear the car is to be left in. At each stage of redescription a deliberate decision should be made on whether further breakdown is necessary.

This decision on further breakdown is a second key principle of HTA. The level of detail of the analysis is determined by the nature of the task and the context in which it occurs. To make the approach more systematic it is recommended that explicit rules are chosen that can be repeatedly applied each time a decision on redescription is needed. One rule that has been used asks analysts to consider both the costs to the system of inadequate performance and the probability of inadequate performance. Thus a critical task with high cost implications, in terms of money or human safety, will be further broken down. On the other hand, a simple task with no cost implications, that can be picked up in the job situation, does not need further detail. As the analysis proceeds, a hierarchical diagram emerges that is not necessarily symmetrical or 'tidy', but is economical in its coverage and contains only the information needed for the purpose in hand.

DIAGRAMS

Figure 23.1 illustrates the hierarchical diagram for an everyday task of operating an overhead projector. There are a number of things to note here. The first is that only a limited amount of information can be recorded in the boxes containing the name of the operation. This reveals the need for some form of supplementary information, typically a table, cross-referenced to the hierarchical diagram. This will be discussed later. If an operation is not further broken down this can also be indicated on the diagram by an underlining of its box. A second feature to note is the presence of 'plans' on this diagram. This is a third defining feature of HTA and now requires a fuller description.

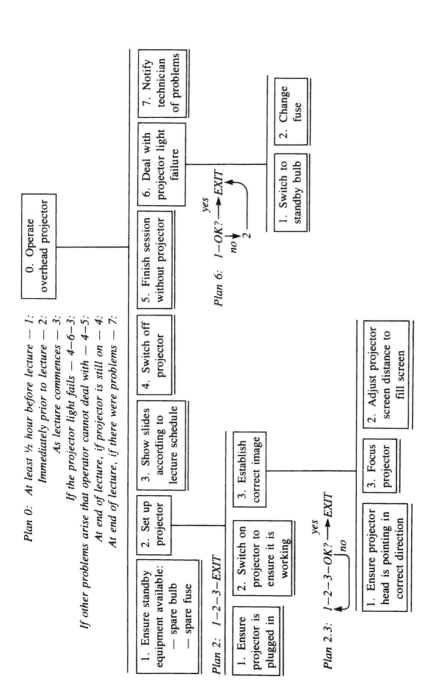

Figure 23.1 Hierarchical diagram with plans (from Shepherd, 1989).

PLANS

When an operation is broken down into two or more sub-operations, there is a need to specify how those sub-operations are organized in relation to each other. For example, should A always follow B? Alternatively, do A and B need to be time-shared or closely coordinated in some way? Another form of organization is where, following a particular outcome of A being carried out, then either B or C should be executed. This is a branching structure. A final type of organization is where either A or B is carried out depending on the prevailing circumstances.

These different forms of organization are termed 'plans' in HTA and are a key to capturing the essence of a task and how it is executed. It becomes important therefore, that as the analysis continues, and operations are recorded, so too are the plans. These plans should take the form of statements on how the sub-operations are ordered. For example, plan 0 in Figure 23.1 specifies a sequence for preparing to use a projector, and then the order of operations in use and at completion of use. Sometimes a diagrammatic version of the plan is more usable. Figure 23.1 presents some simple examples of these. Under some circumstances there can be some complex branching plans to record, and this may be best done with separate diagrams. Plans should also be recorded in tables.

The topic of plans is an often misunderstood area of HTA. It has been thought by some that the approach was only useful for manual sequential tasks, and that the sequence implied in the diagram, reading from left to right, was an implicit plan. It is only by understanding the role of the plan that the key flexibility of HTA can be realized. It was originally designed to capture the detail of complex cognitive tasks where the essence of the complexity lies in the plan. The problem becomes one of deciding what the optimum plan for a task should be and then representing it.

TABLES

It was pointed out above that the HTA diagram will need to be supplemented by more detail and that this was often best done by a table. Various versions of tables have been produced over the years, and from what has been said earlier it should be clear that no 'standard' form of table is specified. This is an area where the purpose of the analysis is likely to be very influential in determining the type of information recorded. The HTA diagram is likely to be fairly consistent across applications, serving as a summary of the task and its structure. It will also serve as an index to the table and a numbering system can be chosen to allow for this. This numbering system can also be used to indicate the relationship of the plans to the operations.

In Figure 23.2 one type of table is illustrated. In this the number of columns is kept to a minimum, different types of information being indicated by typeface and numbering. This type of table has been used in training and HCI applications. It can be seen that there is an initial statement of the first operation. This is followed by the plan in italic type. If there was a complex plan, cross-reference to another diagram could be made. The plan is followed by a list of the sub-operations. If they

Superordinate	Task analysis — operations — plans	Notes

0.	Operate overhead projector
	Plan 0:
	At least ½ hour before lecture — 1;
	Immediately prior to lecture — 2;
	As lecture commences — 3;
	If the projector light fails — 4–6–5;
	If no other problem occurs that you
	    cannot deal with — 4–5;
	At end of lecture — 4 (if on — 7
	    (if problem)) — EXIT	

	1. Ensure standby equipment available — spare bulb — spare fuse	Get replacements from the technician.
	2. Set up projector	
	3. Show slides according to lecture schedule	
	4. Switch off projector	
	5. Finish session without projector	This should never occur. Unfortunately, it sometimes does. Be prepared!
	6. Deal with projector light failure	This is the only fault you should try to deal with yourself.
	7. Notify technician of problems	Failure to do this may cause problems for colleagues or yourself later if equipment is unprepared.

Figure 23.2 A version of a task analysis table (from Shepherd, 1989).

are followed by '//', this means there is no further breakdown of the operation. The absence of that marker means that there is an entry later in the table for that operation. There is a third column in this table where notes can be made for each item. A training analysis would have training comment here, but other types of analysis would adapt this column to their needs.

A second type of table is illustrated in Figure 23.3. This analysis was to be used in a context of selecting types of display for process control tasks. It is not necessary to go into detail here of what the various columns were used for; the intention is to show that the table can be adapted for the purpose in hand. Taking the HTA diagram as a starting point, a table should be produced to represent the information collected in an efficient and usable format. The table should be convenient for the analyst, but what is more important it should be seen as a source document for a design team. They will need a common understanding of the documentation, which should meet all their needs.

Superordinate	Plan	Operations	Information flow across interface	Information assumed	Task classification	Notes
0 Operate coal preparation plant	1 → 2 & 3 until 4. Do 5 as appropriate and 6 at end of shift	1. Start up plant	→ initiate start ← plant items selected	start up procedure	procedural	
		2. Run plant normally	← plant operation & monitoring → control information	knowledge of plant flows & operational procedures	operational	
		3. Carry out fault detection and diagnosis	← fault data	some understanding of faults	fault detection fault diagnosis	
		4. Shut down plant	→ initiate shut down	shut down procedures	procedural	
		5. Operate telephone and tannoy	—	operational knowledge	operational	
		6. Make daily reports	← plant data for log	reporting procedure	procedural	

Figure 23.3   An alternative task analysis table showing information flow, and other types of information for a display design project.

HTA IN USABILITY EVALUATION

Given its general applicability in the human factors world, HTA has a number of roles in usability evaluation. In advance of any system design it can be used to specify user tasks in a task synthesis role. In this case information must be collected from earlier systems and then incorporated with a proposed design to generate alternative configurations of tasks for assessment. Here the hierarchical diagram is a key document. The table can then be used in a different role, where, for example, different ways of carrying the task out can be outlined. It would also be possible to specify different types of interface styles for different operations. The illustration in Figure 23.3 would be an example of this. In a study of a number of industrial plants for coal preparation, Astley and Stammers (1987) classified sub-tasks on the basis of a category scheme developed for that context. Examples of the categories were: monitoring, prediction and problem-solving. These task groupings were then used to choose the optimum type of display for the task in question.

For evaluation activities, HTA can be used to specify typical sequences of operations to be observed; the table entries could include target times for each operation, and could be used to build up a set of task scenarios. These can then be used to assess designs (Stammers and Shepherd, 1995). By looking for common elements across the analysis it should be possible to isolate representative scenarios. Additionally, by using the hierarchy of tasks, it should be possible for a successful test at one level of description to eliminate the need to test at more detailed levels. However, a failure at a higher level can be followed up by examination of the subordinate tasks.

Finally, HTA can be used for the generation of documentation and/or training material for the tasks (Stammers and Shepherd, 1995). The breakdown of the task for HTA forms a clear basis for training programme components. It is also relevant for the sequencing of material to be used in documentation. The plans generated are usefully isolated, and must be included in training and documentation. Scenarios of the kind described above can also be used in training simulations.

HTA sets out to be an economical and flexible approach to collecting and representing task information. This can be for existing tasks or for planned systems where it can be used in a task synthesis mode. Its wide use for different roles and in different contexts commends it for further use. This approach has evolved over the years; it is therefore incumbent on those who use HTA, and make improvements to it, to share this experience by way of appropriate publications.

REFERENCES

ANNETT, J. and DUNCAN, K.D. (1967) Task analysis and training design, *Occupational Psychology*, **41**: 211–221.

ASTLEY, J.A. and STAMMERS, R.B. (1987) Adapting hierarchical task analysis for user–system interface design, in: J.R. Wilson, E.N. Corlett and I. Manenica (Eds), *New Methods in Applied Ergonomics*, pp. 175–184, London: Taylor & Francis.

KIRWAN, B. and AINSWORTH, L.K. (Eds) (1992) *A Guide to Task Analysis*, London: Taylor & Francis.

PISO, E. (1981) Task analysis for process control tasks: The method of Annett *et al.* applied, *Occupational Psychology*, **54**: 247–254.

SHEPHERD, A. (1989) Analysis and training in information technology tasks, in: D. Diaper (Ed.), *Task Analysis for Human–Computer Interaction*, pp. 15–55, Chichester: Ellis Horwood.

SHEPHERD, A. (1993) An approach to information requirements specification for process control tasks, *Ergonomics*, **36**: 1425–1437.

STAMMERS, R.B. and SHEPHERD, A. (1995) Task analysis, in: J.R. Wilson and E.N. Corlett (Eds), *Evaluation of Human Work*, London: Taylor & Francis.

CHAPTER TWENTY FOUR

Task analysis for error identification: applying HEI to product design and evaluation

NEVILLE STANTON

Department of Psychology, University of Southampton, Southampton, UK

CHRIS BABER

School of Manufacturing and Mechanical Engineering, University of Birmingham, Birmingham, UK

INTRODUCTION

Perhaps the reason why studies of human error have not made as much impact as they might is due to the way in which society views the subject matter. People often blame themselves, or others, for error as everyday experience suggests that individuals see shortcomings in themselves when they commit errors. In the short term it is often cheaper to blame the human part of the system (e.g. it is easier to suggest that the user is stupid rather than redesign the machines). In the long term this perspective trivializes the problems and frustrations people experience in their interaction with artefacts (Reason, 1990).

Human error identification (HEI) techniques can be used successfully to alleviate difficulties experienced by users by indicating to the designer when errors are likely to occur in the operation of the device. HEI techniques have typically been reserved for the analysis of complex, high-risk systems (e.g. the operation of nuclear power plants). In this chapter we argue that the use of HEI can be applied to product design and evaluation.

TASK ANALYSIS FOR ERROR IDENTIFICATION (TAFEI)

TAFEI (Stanton and Baber, 1991, 1993, 1995; Baber and Stanton, 1991, 1992, 1994, 1995) differs from most other error prediction techniques by explicitly

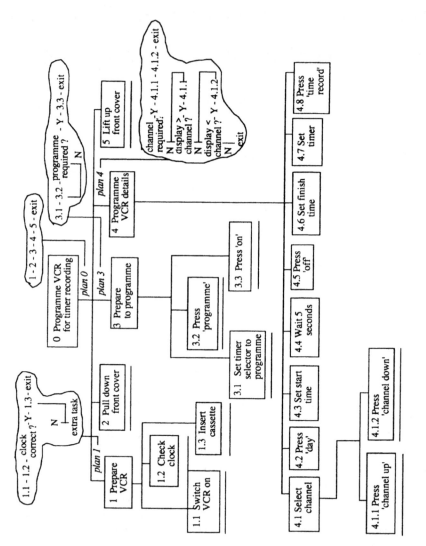

Figure 24.1   Hierarchical task analysis.

analyzing the *interaction* between people and machines. This is done by mapping human activity onto machine states. TAFEI analysis consists of three fundamental components: hierarchical task analysis (HTA), state–space diagrams (SSDs) and transition matrices (TM). HTA provides a description of human activity, SSDs provide a description of machine activity, and TM provides a mechanism for determining potential erroneous activity through the interaction of the human and the device.

Hierarchical task analysis (HTA)

HTA (see Stammers, Chapter 23) is based upon the notion that task performance can be expressed in terms of a hierarchy of goals (what the person is seeking to achieve), operations (the activities executed to achieve the goals) and plans (the sequence in which the operations are executed). The hierarchical structure of the analysis enables the analyst to progressively redescribe the activity in greater degrees of detail (see Figure 24.1). The analysis begins with an overall goal of the task, e.g. '0: Programme VCR for timer recording'. This goal is then broken down into subordinate goals, e.g. '1: Prepare VCR', '2: Pull down front cover', '3: Prepare to programme', '4: Enter programme details' and '5: Lift up front cover'. At this point, plans are introduced to indicate in which sequence the activities are performed, e.g. '1 then 2 then 3 then 4 then 5'. When the analyst is satisfied that this level of analysis is satisfactory, the next level may be scrutinized. The analysis proceeds downwards until the stopping rule ($P \times C$: 'The probability of getting the activity wrong multiplied by the cost of getting the activity wrong') is at an acceptable level. An illustration of a completed analysis is shown in Figure 24.1. The hierarchical diagram provides a useful overview of the task, but does not inform the analyst about the type of interaction with the device. For this to occur, a description of the device is required. TAFEI uses state–space diagrams to describe the different states a device can be in.

State–space diagrams

SSDs are loosely based on finite state machines (Angel and Bekey, 1968). They represent the operational states of the device brought about by human or machine action. Each state is expressed in terms of: a state number, the operational state and the waiting state(s). For example, in state 1, the operational state is 'VCR off'. This has one waiting state: 'waiting to be switched on'. Depending upon the actions of the person operating the device, the VCR will be transformed to a different operational state: 'VCR on'. The SSD is turned into a TAFEI diagram by adding the human activities from the HTA (see Figure 24.2). This serves as a basis for analysis using the transition matrix.

Transition matrices

The transition matrix combines the data from the HTA and SSD to consider each

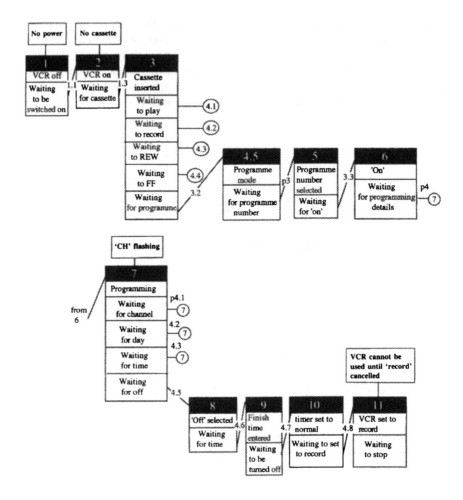

Figure 24.2 State–space diagram.

possible state transition (i.e. the link between each state in the SSD). If a state transition is permitted, then a judgement of whether it is desirable is required. On this basis, each cell in the transition matrix is classified in terms of:

- Legal (L: permitted by the device and desirable)
- Impossible (−: not permitted by the device)
- Illegal (I: permitted by the device but has some undesirable consequences).

An example of these classifications is illustrated in Figure 24.3. For this example, there are 13 transitions defined as 'illegal' (e.g. inserting a cassette into the

To
state:

	1	2	3	4.5	5	6	7	8	9	10	11
1	-	L	I	-	-	-	-	-	-	-	-
2	L	-	L	I	-	-	-	-	-	-	-
3	L	-	-	L	-	-	-	-	-	-	-
4.5	I	-	-	-	L	I	-	-	-	-	-
5	I	-	-	-	-	L	I	I	-	-	-
6	I	-	-	-	-	-	L	I	-	-	-
7	I	-	-	-	-	-	**L**	L	-	-	-
8	I	-	-	-	-	-	-	-	L	-	-
9	I	-	-	-	-	-	-	-	-	L	-
10	I	-	-	-	-	-	-	-	-	-	L
11	-	-	-	-	-	-	-	-	-	-	-

From
state:

Figure 24.3 Transition matrix.

machine when it is off, or switching the machine off in the middle of a programming sequence). In addition, 1 legal transition has been highlighted (at state 7) because it requires a recursive activity to be performed. These predictions then serve as a basis for the designer to address the redesign of the VCR (see Figure 24.4).

Summary of procedure

The flow diagram illustrated in Figure 24.5 indicates the TAFEI procedure. First, the system to be addressed needs to be defined. Next, the human activities and machine states are described in separate analyses. The basic building blocks are hierarchical tasks analysis (for describing human activity; see Figure 24.1) and state-space diagrams. These methods were chosen as they are of long standing, and thus have been proven to work in the field. These two types of analysis are then combined to produce the TAFEI description of human–machine interaction (see Figure 24.2). It is worth pointing out that the state-space diagram also has the potential to contain information about hazards or by-products associated with particular states. From the TAFEI diagram, a transition matrix (see Figure 24.3) is

Transition	Type	Description
4.5-1	1	Switch VCR off
5-1	1	Switch VCR off
6-1	1	Switch VCR off
7-1	1	Switch VCR off
8-1	1	Switch VCR off
9-1	1	Switch VCR off
10-1	1	Switch VCR off
1-3	2	Insert cassette into machine when switched off
2-4.5	3	Programme without cassette inserted
4.5-6	4	Fail to select programme number
5-7	5	Fail to wait for programme "on" light
5-8	6	Fail to enter programme details
6-8	6	Fail to enter programme details
7-7	7	Recursion: some details may be omitted

Figure 24.4 Errors predicted by TAFEI.

compiled and each transition is scrutinized. Each transition is classified as 'impossible', 'illegal' or 'legal', until all of the transitions have been analyzed. Finally, the illegal transitions are addressed as potential errors. These are addressed in turn, to consider changes that may be introduced.

We have found that the use of SSDs as a central concept to TAFEI has had unexpected benefits. These benefits surround the acceptability of the approach by design engineers. In effect we are communicating to them in their own terms: they use the notion of system states on a daily basis, and find that TAFEI is able to represent this formalism in considering human–machine interaction.

TAFEI IN PRODUCT DESIGN AND EVALUATION

TAFEI can be applied in all stages of the product design life cycle. The question is: what degree of efficacy can be achieved in the different stages. This will be considered briefly. In *conceptual* stages of design where the product does not exist, TAFEI can be used to map out the user's goals and the system states. This paper-based exercise could provide a focus for designers to consider the form that the product interaction might take. Once a problem is identified, it is relatively easy to sketch out revised TAFEI diagrams. TAFEI has been used in this manner to consider the design of advanced cruise control systems in cars of the future.

TAFEI can also form part of the *reiterative* design procedure, for example during the rapid prototyping of a product. If the product is not fully functional, some aspects of its use may not be open to observation and a representative range of users interacting with the product may not be feasible. TAFEI can support product development by providing supplementary information regarding the type of interaction and a prediction of errors. We have used the method in this manner to specify the human interaction with a graphical user interface for a medical imaging product.

Finally, TAFEI may be used for *evaluation* of existing products, for example, where extensive observation would be too costly and time-consuming. In a recent study by Baber and Stanton (1995) TAFEI was used to evaluate a ticket vending machine used in the London Underground. This evaluation took only 3.5 hours, compared with direct observation time of 31.5 hours, and yet produced some 80% of the errors that were observed. This indicates that TAFEI appears to have a respectable predictive validity at a fraction of the investment of observation time.

CONCLUSIONS

TAFEI has been used in the evaluation of a wide variety of applications, such as word processors, cash points, VCRs, in-car products, a drain suction truck, and ticket vending machines. We are pleased with the product of these studies, and aim to build up more cases and to encourage others to use the method.

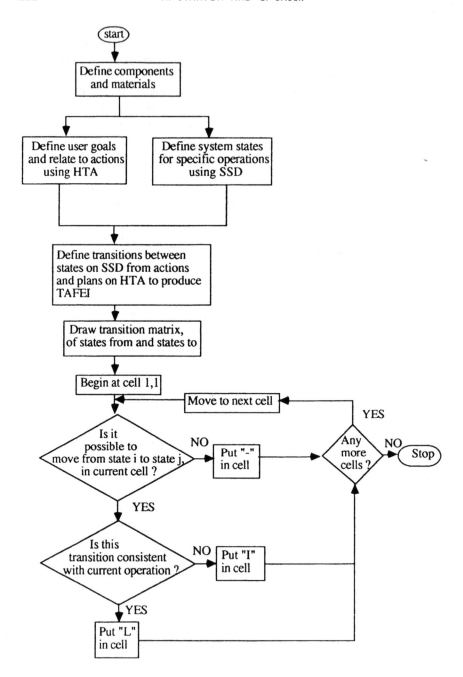

Figure 24.5   The TAFEI procedure.

A comparison of TAFEI with another error prediction method (PHEA; Embrey, 1992) using the criteria developed by Kirwan (1992) shows that the performance of the two appears to be remarkably similar, despite the differences in approach (Baber and Stanton, 1995). Both techniques offer substantial time savings compared with observation, with a small penalty in terms of accuracy. As mentioned earlier, the time taken to perform the HTA and TAFEI analysis took 3 hours 30 minutes in total, compared with over 31 hours spent in direct observation. This suggests that TAFEI can offer a saving of approximately 10:1. We have not yet collected any data on the learning time but, assuming analysts are familiar with HTA and SSDs, we feel that this will be minimal.

In their book *Human Safety and Risk Management*, Glendon and McKenna (1994) conclude that 'The TAFEI technique can thus provide a useful picture of interactions between human operators and machine components within a system in respect of possible actions and errors. From the information gleaned from such an analysis, possible errors which are critical to the safe and efficient operation of the system may be tackled ...'. As with all methods, TAFEI is appropriate for measuring only some of the aspects of usability (e.g. task match and errors). Therefore, it is important to combine TAFEI with other methods when performing a usability evaluation. However, there are not many approaches that can claim to evaluate a product before it is brought into physical existence; this is one of the major benefits of TAFEI.

REFERENCES

ANGEL, E.S. and BEKEY, G.A. (1968) Adaptive finite-state models of manual control systems. *IEEE Trans. Man–Machine Systems*, March, 15–29.

BABER, C. and STANTON, N.A. (1991) Task analysis for error identification: Towards a methodology for identifying human error, in: E.J. Lovesey (Ed.), *Contemporary Ergonomics*, London: Taylor & Francis.

BABER, C. and STANTON, N.A. (1992) Defining problems in VCR use, in: E.J. Lovesey (Ed.), *Contemporary Ergonomics*, London: Taylor & Francis.

BABER, C. and STANTON, N.A. (1995) Comparison of observational data with two human error identification techniques for product assessment: The evaluation of a ticket vending machine, *Applied Ergonomics*.

BABER, C. and STANTON, N.A. (1995) Task analysis for error identification: A methodology for designing error tolerant consumer products, *Ergonomics*, 37(11): 1923–1941.

EMBREY, D.E. (1992) Quantitative and qualitative prediction of human error in safety assessments, *I. Chem. E. Symposium Series No. 130*. London: I.Chem.E.

GLENDON, A.I. and McKENNA, E.F. (1994) *Human Safety and Risk Management*, London: Chapman & Hall.

KIRWAN, B. (1992) Human error identification in human reliability assessment, Part 2: Detailed comparison of techniques, *Applied Ergonomics*, 23: 371–381.

REASON, J. (1990) *Human Error*, Cambridge: Cambridge University Press.

STANTON, N.A. and BABER, C. (1991) *A Comparison of Two Word Processors Using Task Analysis For Error Identification*. Paper presented at the HCI'91 annual conference at

Herriot-Watt University, Edinburgh.

STANTON, N.A. and BABER, C. (1993) *Task Analysis For Error Identification*. Paper presented to the joint MCR/RNPRC/APRC Workshop on Task Analysis, 5–6 April, University of Warwick.

STANTON, N.A. and BABER, C. (1995) Task analysis for error identification: A methodology for reducing human error by machine design, *J. Health & Safety*.

PART EIGHT

Issues Relating to Usability Evaluation

CHAPTER TWENTY FIVE

Performance measurement and ecological validity

MILES MACLEOD

Andersen Consulting, London, UK

INTRODUCTION

Usability evaluation can mean rather different things to different people. Evaluations of the usability of information systems and interactive technology can focus on narrow or broad issues. They may assess conformance with guidelines, reflect expert opinion, or – the concern of this chapter – assess performance in the hands of users. Their scope may range from study of individual interaction with detailed aspects of a user interface, to evaluating the use of a system to support work in an organizational setting. The style, timing, scale and design of any study should be shaped by its objectives, as should the choice of people involved. In practice, it seems that the nature of an evaluation may in some cases be dictated by the immediately available resources, the preferences of staff who are familiar with particular ways of doing such things, or lack of awareness from management of how best to meet their objectives.

This may have contributed to the mixed reputation which usability laboratory studies and associated performance measurement gained in the 1980s. They produced results which too often bore little relation to the quality of system use in the workplace. Evaluations tended to focus mainly on micro-issues, or paid insufficient attention to the characteristics of the people for whom a system was designed, to their work tasks or the work setting. Detailed statistical analyses could lend a gloss of credibility to the results, but managers were often unimpressed. Even valid findings frequently failed to influence design appropriately – a problem familiar in human factors work (Wasserman, 1991) – or simply came too late.

More recently, usability laboratories have increased in popularity, either in their traditional 'television studio' style, or as portable facilities which can be set up in the workplace, or as a custom-built facility which simulates a workplace. Superficially impressive facilities now often provide a focus for drawing attention to usability evaluation, but care is required in their use if misleading results are to be avoided. Work at the UK National Physical Laboratory (NPL) over the past three years, with a range of industrial organizations, has developed practical approaches to usability evaluation – and particularly performance measurement – which explicitly relate method to objectives, and which pay careful consideration to ecological validity: real users, real work goals and realistic settings. The overall approach involves from early in development the participation of a range of stakeholders, including developers, users, trainers, change management staff and quality managers. After early usability input into design, performance is evaluated with prototype systems in realistic work settings, and much of the evaluation is conducted by stakeholders trained in the method, with guidance from usability professionals.

FROM WIDGETS TO WORKFLOWS

Micro-issues are of course important. Successful interaction with interface objects is essential for effective and satisfying interaction between an individual user and a graphically interfaced computer system or an interactive product. This a valid topic for research, and has a long and continuing history, as the proceedings of the most recent human–computer interaction (HCI) conference demonstrate. Experience of evaluating usability shows that even quite small changes in presentation and feedback – both in the design of interaction objects and the manner of their incorporation into dialogues which support work tasks – can greatly influence the quality of interaction. The basic principles for successful design of specific interaction objects are quite well established. They are incorporated into style guides such as the Apple *Human Interface Guidelines* (Apple Computer, 1987), which also emphasize the critical importance of the manner in which such objects are employed in a design. Such guides make it explicit that good design depends upon more than simply using effective widgets; application developers should both follow the general and specific principles, and seek to shape the design to support the achievement by users of their goals.

Design and evaluation are closely coupled at all levels, from the informal testing of design ideas to the formal testing of prototypes and implemented systems. What is at issue here is how best to assess the quality of use of a system for its actual or intended users, with their specific skills and knowledge, and after training if appropriate. But an evaluation which employs unrepresentative users, or atypical tasks, or settings unrelated to the workplace, is unlikely to arrive at findings which bear much relation to the final quality of use of the system being tested.

Many evaluations – and many research studies in HCI and ergonomics – have focused on how individual users interact with a user interface or interactive product. But users rarely work in isolation interacting solely with a computer, especially in organizations. People who use computers to support their work may at the same time also interact with peripheral devices (printers, card readers, etc.), with help systems and user support, with other users on a network, and with colleagues or customers either face-to-face or via the telephone. The balance and nature of these interactions is task-dependent; it varies from workflow to workflow. In short, evaluations of the performance of a system or product in the hands of its users must reflect the wider context in which the system will be used. It is also essential to evaluate how successfully task goals are achieved, both in terms of quantity and quality, as well as the cost of achieving those goals, which may be measured in terms of time or money. These prerequisites underlie the NPL Performance Measurement Method, described later in this chapter.

MEASURING PERFORMANCE

Where information systems or interactive products are designed specifically to support particular transactions and workflows (whether or not re-engineered), a primary concern is their effect on the efficiency of the work they support. Contracts for the development of bespoke systems may incorporate targets for efficiency (or productivity) of work with the new system, expressed in terms of times to achieve specific workflows when the new system is rolled out. Management typically is seeking enhanced productivity and lower staffing levels for tasks which can be computer-supported. Most procuring organizations are also concerned with users' perceptions and staff satisfaction, and some may incorporate targets for 'ease of use' or user satisfaction. Yet targets are only meaningful when they can be expressed quantitatively; it must be possible to measure the efficiency (or productivity) of existing and new work practices, and to measure user satisfaction.

Established work measurement methods typically involve collecting detailed measures of time to perform elements of work (specific activities) gained by observing and timing the performance of work by hundreds or even thousands of staff. Large organizations may have specific work measurement teams, carrying out long-term studies with the intention of informing management decisions about anticipated future staffing level requirements. The individual activities timed may be of only a few seconds' duration. Measures relevant to office information systems often focus on clerical activities. Some methods draw on libraries of data about timings of common clerical tasks, collected from various organizations.

When applied to novel ways of performing work tasks, this approach is essentially bottom-up. It enables estimations of times for composite tasks to be arrived at by adding together the timings for their component activities, often assuming an idealized way of carrying out the task. This kind of work measurement has some commonalities with the Keystroke-Level Model (K-LM; Card et al., 1980) for estimating task performance times for interaction between individuals and computer systems, which assumes error-free expert performance (although the K-LM also incorporates an explicit model of the user's cognitive processing time, represented by the 'M' operator). Clerical work measurement methods typically do not make explicit allowance for cognitive processes. In some cases they may add a time element for error recovery. However, both these factors may simply be reflected in the mean performance times for some activities.

These traditional approaches to performance measurement may in some circumstances give useful approximations of performance times, but they have the clear disadvantage of relying on historical performance data to draw inferences about work performance with as yet unimplemented systems, which will be affected by different contextual factors. The productivity of computer-supported work is affected by specific design decisions at every level, and where new systems are being developed these traditional methods do little to help shape many key design decisions which will affect efficiency and user satisfaction.

Many organizations keen to improve productivity appreciate the critical role of staff training in enhancing the efficient performance of work tasks. However, assessment of the effectiveness of training too often takes the form of tests of knowledge (people's ability to recall correctly, or at least recognize facts or procedures in which they have been trained), rather than assessment of the effects of training on performance. When evaluating usability, the amount and nature of the training which users have received at the time of the evaluation can have major effects on both the measured efficiency and user satisfaction. In the approach advocated here, user training and support materials are iteratively developed and improved as part of the usability engineering process.

BRINGING TOGETHER METHODS AND PEOPLE

The requirements are challenging but clear. There is a need for practical and cost-effective usability evaluation and performance measurement methods which are ecologically valid (i.e. which reflect relevant aspects of the work context); which can be carried out before implementation of a system; which provide diagnostic data to help shape a prototype system to meet user and organizational needs; which win management support and involve key stakeholders, and which feed into the development of training and user support materials.

The approach developed by NPL for meeting these requirements draws on outputs of the ESPRIT project MUSiC (Measuring Usability of Systems in Context), especially Usability Context Analysis (UCA; Maissel et al., 1993); the NPL Performance Measurement Method (Rengger et al., 1993), supported by video analysis software; and SUMI, the Software Usability Measurement Inventory (Porteous et al., 1993; see elsewhere in this volume). The approach also draws on the emerging ISO Standard, 9241-11, *Guidance on Usability* (ISO, 1994). ISO 9241-11 views usability in terms of quality of use, defined as the effectiveness, efficiency and satisfaction with which specified users achieve specified goals in particular environments. These inputs provide the raw materials, but much of the value has been achieved by refinement through practical application in commercial evaluations. The approach employs realistic work scenarios, identified by UCA, and credible simulations of the workplace conditions.

Whenever possible, the usability engineering process should commence early in the development of an information system, and the system design should be refined through iterative prototyping. The NPL Usability Services approach brings together a combination of in-house and external skills. It has been developed to enable organizations to employ their own staff to carry out valid usability evaluations and measurement, applying MUSiC and other evaluation methods, with limited but sufficient specialist input.

After winning management support, the first step is to identify stakeholders, and to encourage the formation in the client organization of a small, multi-interest usability team to guide what should be evaluated. As a minimum this should include someone with human factors knowledge, people representing user interests and training and support material development, and someone from the product development team. These people inform and maintain contact with other stakeholders, and are well placed to influence the feeding back of results into system development, into work practices and into user training. The in-house usability team after training in usability measurement carries out the evaluation work, with some specialist input. Management of usability work requires both *knowledge of usability issues* and *the ability to influence decisions* in parts of the project which conventionally may not work in close liaison.

USABILITY CONTEXT ANALYSIS (UCA)

UCA, initially developed in MUSiC by NPL and the HUSAT Research Institute, is a practical, structured method for identifying and documenting in simple terms the characteristics of the information system's (or product's) users, the tasks it is intended to support, and the anticipated circumstances of system use. Contextual factors which may have potential to affect usability are described at a fairly broad level, under a number of headings (now incorporated into ISO 9241-11), and are together referred to as the 'context of use'. The method is supported by a context questionnaire which gives guidance on what to consider in answering each question. Subsequent steps in UCA involve identifying which characteristics are potentially relevant to the usability of the system being evaluated; defining the context of evaluation; and producing a concise evaluation plan. UCA thus helps to ensure that usability evaluations reflect the context of use and give data with acceptable ecological validity. UCA forms an integral first component of the NPL Performance Measurement Method.

UCA can be applied early in system development or redesign, and is most effective when carried out cooperatively with a selected range of stakeholders. This raises awareness of usability factors, enables effective and timely communication about user needs and usability, and facilitates agreement and shared understanding on user-centred issues. It provides a vehicle for airing possibly differing views, and arriving at an agreed view of those issues which are relevant to quality of system use. The formal work of UCA in the NPL approach centres on one or more facilitated group meetings. Between six and twelve relevant key stakeholders should participate. They must be given adequate briefing in advance about the aims of UCA, and the information they should bring to the meeting. Some typical areas from which stakeholders can be identified are shown in Figure 25.1. Further areas

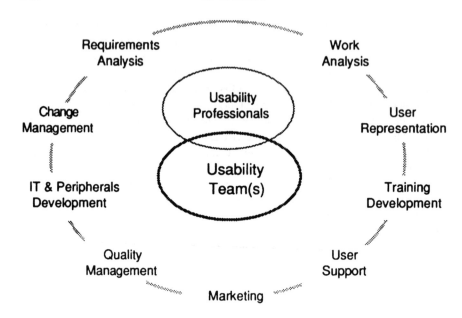

Figure 25.1 Interest groups from which stakeholders are identified, and the core usability team is formed.

include hardware and furniture provision, and health and safety – especially relevant in the light of national legislation implementing Directive EC 90/270 on health and safety of work with VDUs. The group should include people with sufficient seniority to make or influence decisions concerning the evaluation of usability of the product and the course of its development.

In defining the context of use, the aim is to arrive at a fair summary of the contextual factors relevant to use when a system is implemented, and *not* to consider the constraints or circumstances of system evaluation. Once agreed – with facilitation from the person leading the context meeting – each answer is recorded in the UCA report table. The aim of the subsequent steps is to ensure that evaluation is carried out in a context which fairly reflects real-life system use. Each documented factor of the context of use is considered, and its relevance to usability is assessed. This requires some knowledge of human factors or HCI, and is best carried out separately from recording the context of use. For each contextual factor identified as possibly affecting usability, the usability analyst (in consultation with members of the usability team) must decide how to control or monitor that factor in the evaluation. Decisions are recorded in the UCA report table, to summarize the context of evaluation. This specifies the conditions under which the evaluation should take place. An evaluation plan can then be derived describing the practical details of how the evaluation will be conducted.

Key issues to be decided in shaping the evaluation plan are the choice of the evaluation tasks and users. Frequent and critical tasks relevant to the specific evaluation goals are usually selected, in liaison with key stakeholders, although task choice may be constrained by the functionality of the available prototype and peripherals, and available training and user support. In identifying the profiles of users for an evaluation it is wise to allow some flexibility, and to indicate the desirable spread or homogeneity of user characteristics, because of inevitable constraints on availability of suitable users within the resources and timescale of the evaluation. The difficulties of locating appropriate users are well known. UCA offers two advantages. First, it provides advance notice of the characteristics of users required for an evaluation. It may be necessary, for example, to give users training in the use of a prototype system. Involving a training manager in UCA helps in the early setting up of an appropriate training programme, both for the evaluation and for use of the implemented system. Secondly, UCA enables subsequent interpretation of evaluation findings to be informed by knowledge of how well the users employed in an evaluation matched the profile of intended users of the system, in terms of those characteristics identified as being relevant to usability.

THE NPL PERFORMANCE MEASUREMENT METHOD

Many approaches to performance measurement simply deliver times for task completion. Some deliver counts of 'errors'. In evaluating usability and measuring performance, the NPL method focuses on the *quality* and *degree* of work goal achievement, as well as measuring its cost in terms of task times, and identifying areas of difficulty. The method gives measures of core indicators of usability identified in ISO 9241-11, which relate directly to productivity and business goals: effectiveness (how correctly and completely goals are achieved in context) and efficiency (which relates effectiveness to cost in terms of time). It also gives measures of productive period (the proportion of time spent not having problems), and of Snag, Search and Help times: time spent overcoming problems, searching unproductively through a system, and seeking help. These problem-related measures are optional components of the method, because they can only be derived by retrospective video analysis. In practice, it is usually considered cost-beneficial to derive them, since they provide valuable sources of diagnostic data, and assist in identifying specific areas of difficulty and providing pointers to causes of problems.

While simple observation can yield rich usability data, analysis solely in real time may lose valuable data. However, observational data recorded on video can be very time-consuming to analyze, conventionally it takes ten or more hours to analyze a single hour of video. This is unacceptable in commercial evaluations. Video recording offers many advantages and preserves data which are not cheap to collect, but there is a real need for efficient support for video analysis, to enable cost-effective derivation of the objective performance measures and diagnostic information. The Diagnostic Recorder for Usability Measurement (DRUM; Macleod and Rengger, 1993) is a software tool which helps reduce analysis time to 2 or 3 hours per hour of recording, and supports first-pass analysis in real time. DRUM provides support for many aspects of usability evaluation. It assists management of data throughout evaluation; task analysis; video control; creation of an interaction log of each evaluation session; analysis of logged data; and derivation of the NPL MUSiC performance measures and metrics, and user-defined measures. DRUM also supports the identification of analyst-defined critical incidents, and diagnosis of specific usability problems. Once logged, any observed event can be automatically found and reviewed on the video. Video clips of specific observations can be assembled into a highlight tape, which provides convincing evidence for managers, developers and trainers of areas in a design which need further attention (or alternatively gives highlights of usability successes).

In applying the NPL method, the diagnostic aspects of user-based evaluations are augmented by outputs from SUMI, a standardized psychometric questionnaire for measuring dimensions of perceived usability (user satisfaction), which also gives rich diagnostic data. Efficiency and user satisfaction are not necessarily correlated (a system can be satisfying but not very efficient to use, or vice versa), so there is a great advantage in measuring both. The SUMI questionnaire is administered at the end of a user session. SUMI takes about five minutes for each user to complete, and is quick to analyze subsequently with the help of the SUMISCO analysis software.

The NPL approach to usability evaluation and getting usability into system development, outlined in this chapter, has been taken up by some major organizations, and has been refined progressively in the light of practical experience. Its successful application depends not just on contextual validity, but on practicality, flexibility, timeliness, and the appropriate use of methods and supporting tools. Above all, successful usability evaluation and performance measurement require the involvement and participation of key stakeholders. Their involvement is essential, especially in facilitating the incorporation of identified improvements into system design.

REFERENCES

APPLE COMPUTER INC. (1987) *Human Interface Guidelines: The Apple DeskTop Interface*, Wokingham, UK: Addison-Wesley.

CARD, S., MORAN, T.P. and NEWELL, A. (1980) The keystroke-level model for user performance time with interactive systems, *Communications of the ACM*, **23**(7): 396–410.

ISO (1994) DIS 9241-11, *Ergonomic Requirements for Office Work with Visual Display Terminals (VDTs)*, Part 11: *Guidance on usability*, International Standards Organization.

MACLEOD, M. and RENGGER, R. (1993) The development of DRUM: A software tool for video-assisted usability evaluation, in: J.L. Alty *et al.* (Eds), *People and Computers VIII* (Proc. HCI'93 Conf., Loughborough, UK, September 1993), pp. 293–309, Cambridge: Cambridge University Press.

MAISSEL, J., MACLEOD, M., THOMAS, C., RENGGER, R., CORCORAN, R., DILLON, A., MAGUIRE, M. and SWEENEY, M. (1993) *Usability Context Analysis: A Practical*

Guide, V3.0, NPL, DITC, Teddington, UK.

PORTEOUS, M., KIRAKOWSKI, J. and CORBETT, M. (1993) *SUMI User Handbook*, Human Factors Research Group, University College Cork, Ireland.

RENGGER, R., MACLEOD, M., BOWDEN, R., DRYNAN, A. and BLAYNEY, M. (1993) *MUSiC Performance Measurement Handbook, V2*, NPL, DITC, Teddington, UK.

WASSERMAN, A.S. (1991) Can research reinvent the corporation? A debate, *Harvard Business Review*, March–April, p.175.

CHAPTER TWENTY SIX

Issues for usability evaluation in industry: seminar discussions

PATRICK W. JORDAN, BRUCE THOMAS and IAN L. McCLELLAND

Philips Corporate Design, 5600 MD Eindhoven, The Netherlands

INTRODUCTION

The aim of this chapter is to reflect the main issues which emerged from these discussions. The issues identified are based on points raised in the discussion by the seminar participants. However, the conclusions drawn are those of the authors of this chapter.

Broadly, there appeared to be three major themes which emerged from the discussions:

- The acceptance and expansion of usability as a concept. This has brought increasing demands on those involved in usability practice – these come both from the customer and from others involved in product manufacture.

- The recognition that usability is a part of ensuring product quality and the integration of usability issues into the design process.

- The contribution that disciplines other than human factors can make to usability issues and the need to forge effective working partnerships.

These themes are expanded in this chapter, after which some overall conclusions are drawn.

USABILITY AS A CONCEPT

The phrase usability probably first came into usage in the late 1970s or early 1980s. The concept was developed further during the 1980s with significant contributions coming from, amongst others, Bennett (1984), Eason (1984) and Shackel (1986). Building on this work, the International Standards Organization is currently developing a formal usability standard (ISO DIS 9241-11). They propose the following definition of usability.

237

> ... effectiveness, efficiency and satisfaction with which specified users achieve specified goals in particular environments.

This standard has provided a useful focus for usability issues. Indeed, this is reflected in the many references to the standard throughout this book. Simply having a standard can prove beneficial in itself. Professionally, it can be seen by others as a formal representation of the issues concerning usability practitioners. Formalizing usability as a concept is helpful in terms of encouraging the inclusion of usability criteria in design specifications and incorporating human factors issues into the design and evaluation process.

From the discussions at the seminar, it appears that many involved in product manufacture are starting to see usability as a central issue in ensuring product quality. As a concept, usability addresses the match between the product and the tasks for which it is intended, the users for which it is intended, and its context of use. This is something that manufacturers are beginning to realize that they must get right in order to gain a competitive edge with their products. The involvement of usability practitioners in the design process, then, represents an acknowledge-ment of the need to address quality at the user interface in a systematic way.

Usability as an issue first achieved prominence through human factors professionals involved in the design and manufacture of computers and computer software. Most of the early work on usability was, then, done in the context of human–computer interaction (HCI), usually relating to professional users and concentrating on the functional performance of products. Now, however, usability issues have been taken up in many other areas. This was reflected in the backgrounds of the participants in the seminar, and the topics covered by the presentations and discussions. Although many of the products under discussion were software based, there was also emphasis on consumer electronics products, where users' attitudes were regarded as being at least as important as functional performance.

The chapters in this book, along with much of the discussion that took place during the seminar confirm and support other evidence (e.g. Nielsen, 1993; Wicklund, 1994) for the take-up of usability issues in industry. In addition to the number of organizations now employing usability specialists, an encouraging trend appears to be the number of these which have equipped on-site laboratories for usability evaluation (again, the discussion at the seminar supported other evidence for this; e.g. Nielsen, 1994). As well as being a demonstration of organizations' commitment to designing for usability, these laboratories can provide a useful focus for the usability-related activities going on. This can have a beneficial 'propaganda' effect, being a useful vehicle for demonstrating usability issues to others. For example, designers or those commissioning design and evaluation work might come along to watch a test in progress. Direct viewing of users interacting with products can prove a convincing way of getting a message across. There is no substitute for seeing users struggling with a product. Certainly, it may have a greater impact than a written report, where the readers may not always be convinced that the problems mentioned by the usability practitioners are really that serious.

Much of the seminar discussion raised issues about how the incorporation of usability into the design process and the evaluation of usability could be achieved as effectively and efficiently as possible. However, it appeared that in the organizations where most of those present worked, the idea that usability was a central tenet of good design was less in question. If such a seminar had been held, say, five years ago, it might have been expected that much of the discussion would have centred on how to convince those commissioning designs that usability needed to be addressed at all. Now, however, the emphasis appears to have shifted to the way in which usability issues are handled – for example, pursuading those commissioning work to commit the appropriate level of resources to tackling usability issues, and to bringing in usability practitioners throughout the whole of the product creation process.

As usability has received more attention and has become incorporated into products for use in a wider range of contexts, the range of issues addressed under the banner of usability has also increased. Earlier usability work tended to concentrate on taking objective measures of performance, such as time on task and error rate, usually from the observation of those with some degree of experience with the product. Although such measures may still give very important information about the adequacy of a product, the benefits of tackling wider issues in the context of usability are also being recognized by practitioners.

A reflection of this was the range of methodology presented and discussed at the seminar. Methods aimed at quantifying objective performance were still in evidence, although many of the methods discussed were aimed at capturing users' attitudes towards products. There was also evidence of a product, or those using a product for the first time. The co-discovery method, for example, can be a useful technique for addressing the difficulties faced by new users.

Within the attitude component of usability, the range of issues addressed has widened. In earlier evaluations, measuring attitudes in the context of usability evaluation usually meant trying to quantify the users' levels of satisfaction with a product, where satisfaction referred to 'comfort and acceptability' (ISO DIS 9241-11). In general terms, then, the importance of avoiding exposing the user to any discomfort was recognized. This has long been seen as a central issue in usability evaluation. For example, in an earlier paper on usability, Shackel (1986) recognized the importance of product use being 'within acceptable levels of human cost in terms of tiredness, discomfort, frustration and personal effort'. The emphasis of usability, then, may have been seen as avoiding negatives – ensuring that interfaces were free of bugs that would inhibit performance or cause the user annoyance or difficulty.

Much of the discussion at the seminar, however, indicated that there appears to be a move to the consideration of how usability can make a positive contribution, in terms of emotional benefits for the users, rather than simply protecting the users from negative feelings. The importance of reflecting users' values in products was recognized. This might include, for example, ensuring that a product's design and manufacture conform to the sort of ecological values that the user finds acceptable. Similarly, designs can reflect, for example, values that might be traditionally associated with, say, masculinity or femininity. A major issue is how to incorporate and evaluate pleasure in product use.

Games provide interesting subject matter in this respect. These may be appealing because they are difficult to use. Musical instruments can also create pleasure in the use of difficult products. Both in games and music, the skill of the user can be an important element in the enjoyment of the product. With professional product use, similar phenomena occur. In some occupations mastery of the technology is a matter of professional pride. It may be, then, that making some types of products too easy to use 'de-skills' the job of those using them, making the tasks of those using them less enjoyable. This might lead to a possible trade-off between usability in terms of functional performance and the emotional aspects of product use.

USABILITY AS PART OF THE DESIGN PROCESS

As manufacturers begin to see quality of use as part of overall product quality, usability issues are becoming more integrated in the design process as a whole. At one time usability may have been seen as an 'add-on'. This meant that often usability issues would not be raised until the final stages of the design. Perhaps the attitude might have been, '...we've designed the product, now let's make sure it is easy to use'. In this context, usability evaluations may have been seen as checks to see if whatever had been done had actually helped to make the product more usable. The outcomes of these might provide useful inputs into future designs, although it might inevitably have been too late to do anything about improving the design that was under consideration at the time. Now, however, there are a variety of methods available for addressing usability issues right from the start of the design process. These include, for example, techniques for the evaluation of user characteristics, requirements capture, and the evaluation of non-working prototypes. The need to manage usability issues throughout the design process emerged as a major outcome of the discussions. Strongly linked to this is the issue of the appropriateness of applying different methods at the various stages of design. Van Vianen *et al.* (Chapter 2) and Stanton and Baber (Chapter 5) outline approaches to this issue.

Perhaps because of this more extensive involvement, there seem to be increasing demands on usability practitioners from others involved in product manufacture, both in terms of the quality and accuracy of the information which they are expected to provide and the speed at which they are expected to provide it. A complaint that has often been levelled against human factors work is that the issues identified in the laboratory as being highly significant (often this means issues that show a statistically significant effect on performance) may not be that important in the 'real-life' context in which the product will actually be used.

A reason commonly cited for this discrepancy – and one which came under discussion at the seminar – is the artificiality of many laboratory environments and evaluation protocols. Laboratories will often be free of the sorts of distractions that users will encounter when using a product in its real context of use and evaluation scenarios often request users to perform tasks in isolation which, in reality, they might do whilst also undertaking other activities – for example, doing other related or unrelated tasks, or simply chatting to colleagues. Equally, the somewhat sterile atmosphere afforded by a laboratory environment may prove intimidating to evaluation participants. This might have a negative effect on performance or change the ways in which users interact with products.

The issue of the extent to which the environmental context in which an evaluation is conducted mirrors the product's real-life context of use is known as ecological validity. Often, there will be a trade-off between ecological validity and the tightness of controls on an evaluation. Because of the many other tasks that might be going on when a product is used in its normal context, it can be difficult to know whether or not a problem in usage is really being caused by the product, or whether it is simply due, for example, to some other distraction that is affecting the user. In the laboratory, the converse is true. Here it may be that effects on performance that appear highly significant when all distractions are removed are rendered unnoticeable when the product is in its normal context of use.

One way in which usability practitioners have tried to tackle the control/validity trade-off is to try to simulate (in the laboratory) the type of environment in which a product would typically be used. This issue of scenario creation arose during the discussions in the seminar. For example, if the product under test were a telephone for use in a business context, it was thought that it might be appropriate to arrange the laboratory to mirror the interior of a typical office. Scenario creation need not be limited to the physical test environment. It is also possible to generate a scenario through the way in which tasks are set. For example, instead of simply setting users a sequence of tasks to do, it may be possible to develop a 'story' – perhaps by asking participants to pretend that they are, say, secretaries working for a particular type of company and presenting the tasks as if they were problems that might arise during their working day. Whilst it was felt that there may well be benefits in creating scenarios (for example, in terms of making the evaluation participants more comfortable and relaxed than they would be in a room that actually looked like a laboratory) it was considered to be highly questionable as to whether ecological validity can really be achieved this way. There would still be a lot of artificiality with such a set-up. However faithfully the real environment is physically reproduced and however appropriate the story, the participant still knows that he or she is involved in a usability evaluation and this is almost certain to affect performance to a certain extent. For example, the participant's motivation for using the product might be very different if they were using it in its normal context.

In order to assist with design decisions throughout product development, those involved in the process often require speedy feedback from usability evaluations. One way in which usability practitioners have addressed this is to develop 'quick-and-dirty' evaluation techniques. Broadly, a quick and dirty evaluation is one in which some elements of an evaluation method are relaxed in order to save time and money. For example, a less than ideal number of users might participate in an evaluation or controls and balances may be relaxed. This will, almost inevitably, be at some expense to the accuracy and quality of the information gained from the evaluation. An entire session of the seminar was dedicated to the discussion of informal usability evaluation techniques. The question most often facing those conducting this type of evaluation was how to maximize the time savings at minimum cost to the integrity and usefulness of the results. This issue is one which has apparently received little attention in terms of being systematically investigated.

PARTNERSHIPS

Cooperation with other professions, particularly with designers, is necessary to ensure the full integration of usability issues throughout the design process.

A key, then, to becoming integrated in the design process is the ability to work in multidisciplinary teams. A role of the human factors specialist in such teams is to provide reflection on design. Usability evaluation can be seen as an opportunity for doing this. The usability practitioner can also provide information about users' requirements and ideas for design directions. In order for human factors issues to be recognized and designed into products, the schism of human factors specialists versus designers must disappear. One delegate commented that designers are not 'beings from another planet'. Working together with designers to develop usable solutions should ultimately be more effective than taking a role solely as an evaluator, as this might make the usability practitioner appear somewhat 'aloof'.

The need to work with other professionals is not limited to designers alone. Others with a responsibility for customer satisfaction are also involved. There is a growing need for cooperation with marketing, sales and product management, and for people in these disciplines to recognize the role that a human factors specialist can play. One simple way suggested of promoting cooperation was to involve product managers as participants in usability evaluations; this can operate as an effective propaganda tool, giving them a feel for what it is like to be on the 'receiving end' of the design decisions that are taken.

Working in teams shows that the disciplines involved in product creation are not wholly separate. In the process of synthesizing solutions many have skills to contribute. In teams, the boundaries between disciplines can be broken down, so not only can human factors specialists provide a creative input to the design, but also others can become involved in areas that are traditionally regarded as being in the human factors domain. For example, the seminar discussed the positive effects of including designers and engineers in user requirements capture.

Ways in which this type of communication could be promoted were discussed. One suggested approach was the development of tools and techniques that could provide a platform for communication between the various groups of professionals. This included the possibility of providing designers with tools for performing their own usability evaluations. Checklists outlining the principles of designing for usability might be one such approach. These sorts of tools can be beneficial in that they bring designers and others into direct contact with usability issues and can thus enhance their understanding of human factors practice. However, it was also felt that there could be a danger that encouraging others to use such tools might have the effect of appearing to 'trivialize' usability issues. Whilst tools such as checklists can be helpful, they often do not explicitly take into account factors such as the context in which a product will be used and the characteristics of those using the products. In any case, incorporating issues such as these into an evaluation can often require specialist judgements which those without a human factors background might not be expected to be able to make.

CONCLUSIONS

The picture of the state of usability evaluation in industry which emerged from the seminar was generally positive. It appears that those involved in product manufacture are beginning to see addressing issues as part of guaranteeing product quality. As a result of this, usability issues are being considered to an unprecedented degree throughout the design process.

With the increased attention given to usability issues have come increased pressures on those responsible for ensuring a product's quality of use. Not only is this so in terms of the widening range of issues that usability covers, but also in terms of the quality and complexity of information required and the speed at which it must be provided.

The next few years are likely to be an exciting and challenging time for usability practitioners. As we approach the end of the century, the discipline appears to be gaining a foothold in the product creation process. Usability practitioners should seize this opportunity to contribute to producing useful, usable products that will delight the customer.

REFERENCES

BENNETT, J.L. (1984) Managing to meet usability requirements: Establishing and meeting software development goals, in: J. Bennett and D. Care (Eds), *Visual Display Terminals*, pp. 161–184.

EASON, K.D. (1984) Towards the experimental study of usability, *Behaviour and Information Technology*, 3(2): 133–145.

ISO (1994) DIS 9241-11, *Ergonomic Requirements for Office Work with Visual Display Terminals (VDTs)*: Part 11: *Framework for describing usability in terms of user-based measures*. International Standards Organization.

NIELSEN, J. (1993) *Usability Engineering*, London: Academic Press.

NIELSEN, J. (1994) *Behaviour and Information Technology* (special issue: Usability Laboratories), 13(1 and 2).

SHACKEL, B. (1986) Ergonomics in design for usability, in: M.D. Harrison and A. Monk (Eds), *People and Computers: Designing for Usability*, pp. 44–64, Cambridge: Cambridge University Press.

WICKLUND, M.E. (1994) *Usability in Practice*, London: Academic Press.

Index

Milton Keynes UK
Ingram Content Group UK Ltd.
UKHW040109071024
449327UK00019B/931